Charles Seale-Hayne Library
University of Plymouth
(01752) 588 588
LibraryandITenquiries@plymouth.ac.uk

Innes M. Keighren was until recently Research Associate
at the Institute of Geography and the Centre for the History of
the Book at the University of Edinburgh. He is now Lecturer in
Human Geography at Royal Holloway, University of London.

Bringing Geography to Book

Ellen Semple and the Reception of Geographical Knowledge

INNES M. KEIGHREN

I.B. TAURIS

LONDON · NEW YORK

Published in 2010 by I.B.Tauris & Co Ltd
6 Salem Road, London W2 4BU
175 Fifth Avenue, New York NY 10010
www.ibtauris.com

Distributed in the United States and Canada Exclusively by Palgrave Macmillan
175 Fifth Avenue, New York NY 10010

Tauris Historical Geography: 4

ISBN: 978 1 84885 141 2

A full CIP record for this book is available from the British Library
A full CIP record is available from the Library of Congress

Library of Congress Catalog Card Number: available

Printed and bound in Great Britain by CPI Antony Rowe, Chippenham
from camera-ready copy typeset by Alan G. Mauro and edited by the author

MIX
Paper from
responsible sources
FSC
www.fsc.org   FSC® C013604

Contents

Figures

Acknowledgements

I acknowledge with grateful thanks the Arts and Humanities Research Council, whose funding permitted the research upon which this book is based. Periods of archival work were made possible by a Small Research Grant from the Royal Scottish Geographical Society; a Small Project Grant from the University of Edinburgh Development Trust; and a Helen and John S. Best Research Fellowship from the American Geographical Society Library and University of Wisconsin-Milwaukee Center for International Education. Karl Raitz and Pradyumna P. Karan were kind enough to arrange for me a Visiting Scholar position at the Department of Geography, University of Kentucky. My time spent in Semple's home state was invaluable, and I am grateful for that opportunity.

This book would not have been possible without the kind assistance and skilful guidance of librarians and archivists at a number of institutions: the American Geographical Society; the American Philosophical Society; Clark University; Harvard University; the National Library of Wales; the Newberry Library; Princeton University; the Royal Geographical Society; the University of Cambridge; the University of California at Berkeley; the University of Chicago; the University of Kentucky; the University of Oxford; and the University of Wisconsin-Milwaukee.

I should like to acknowledge the use of material that has, in various earlier forms, appeared elsewhere in print. Chapters 3 and 5 draw from ideas presented first in a paper published in *Transactions of the Institute of British Geographers* 31/4 (2006), pp. 525–40. Elements of Chapter 4 were developed in an article appearing in the *Scottish Geo-*

graphical Journal 124/2 & 3 (2008), pp. 198–203, and aspects of my discussion in Chapter 5 of the reception of *Influences* in the United States are outlined in M. Ogborn and C. W. J. Withers (eds), *Geographies of the Book* (Farnham: Ashgate, 2010). For permission to incorporate this material, I thank John Wiley & Sons, Taylor & Francis, and Ashgate respectively.

Sincere thanks go to Charles W. J. Withers, who has been unfailingly generous with time and ideas. I am grateful too to Geoffrey J. Martin and William A. Koelsch, who read through an early draft of this book. They put me right and set me straight. Any errors that remain are my own. David N. Livingstone and Lynn A. Staeheli did much to guide my thoughts on scale and reception and I am appreciative of their insights. Robert J. Mayhew and David Stonestreet have been patient and encouraging editors and I thank them for their guidance and enthusiasm. Tracey Learoyd provided invaluable assistance in drafting the map which appears as Figure 5. For assistance in translating German and Italian texts respectively, I am indebted to Michael Burn and Rosa Salzberg. The constant support of my friends has been wonderful and crucial. My thanks to them all.

1   Geography, the book, and the reception of knowledge

On 20 September 1912 some sixty geographers, representatives of more than a dozen nations, gathered in Muir Woods, a grove of giant redwood trees north of San Francisco.[1] These geographers were members of the Transcontinental Excursion – a 13,000-mile, eight-week geographical expedition organized by the Harvard geomorphologist William Morris Davis (1850–1934) to mark the sixtieth anniversary of the American Geographical Society. The stand of lofty redwoods, beneath which the party assembled for a commemorative photograph (Figure 1), was 'a living commentary on the effect of climate, modified by altitude and exposure.'[2] In miniature, as it were, Muir Woods illustrated a contemporary trope in geographical thought: environ-

1 Geographers at Muir Woods, California, during the Transcontinental Excursion of the American Geographical Society, 1912

mentalism. This was not environmentalism as it is understood today (a concern for the preservation and protection of the natural world), but was a modern expression of a Classical theory which held the physical environment to be an explanatory mechanism in biological and social development. For the Excursion's members, the question of environmental influence – of the role of geography in shaping the development not just of redwoods but also human society – was a shared and pressing concern: its 'systematic elucidation' was seen to be a 'great and worthy task.'[3]

For much of the first quarter of the twentieth century, Anglo-American geographical inquiry was characterized by its engagement with environmentalism. Scientific efforts to describe and explain the ways in which environmental circumstances conditioned and constrained human societies unified the discipline's intellectual focus and thus facilitated its academic institutionalization. The principal spur to geography's engagement with what would later be termed, and pejoratively so, "environmental determinism", was the work of Ellen Churchill Semple (1863–1932), and particularly her 1911 volume *Influences of Geographic Environment*.[4] Semple's book, part of a venerable tradition of environmentalist scholarship, built upon the work of German geographer Friedrich Ratzel (1844–1904). *Influences* was an adaptation and restatement of Ratzel's concept of anthropo-geography, a theory that saw the distribution and development of human societies as a function of their physical environment. Revised and expanded by Semple ('misinterpreted' in the view of some), an-thropogeography held that religion, politics, economics, and settle-ment patterns, as well as a population's physical and mental charac-teristics, could be understood by reference to the persistent influence of topography and climate.[5] Anthropogeography was thus promoted as an empirical methodology which would provide geography with a 'scientific foundation' to the study of human affairs.[6]

By outlining an approach to geographical research that satisfied a desire for quantification and scientific rigour, *Influences* 'determined the methodological thought of at least a generation.'[7] For a time, in an Anglo-American context at least, *Influences was* geography and was implicated in the discipline's 'scramble for intellectual turf.'[8] In the opinion of one contemporary, Semple's book shaped 'the whole trend and content of geographic thought.'[9] For the geographers of the

Transcontinental Excursion – nearly all of whom subsequently read *Influences* and incorporated it in various ways in teaching and re-search – environmentalism was a live and urgent concern. Narrated in this progressivist mode, the history of geography's engagement with environmentalism, and the discipline's response to Semple's book, appear benign and uncomplicated. The reality, of course, was rather different.

Despite the undoubted importance of *Influences* in directing the initial course of Anglo-American disciplinary geography, Semple's environmentalist ideas, of which her book was the foremost represen-tation, were not received with uniform enthusiasm. For those geog-raphers who considered *Influences* a monument to Semple's scholar-ship and erudition, her book was a timely manifesto for a scientific and anthropological approach to geographical research. For other geographers, however, Semple's book was conceptually flawed – a text which might damage geography's academic legitimacy and dis-ciplinary credibility. Accepted by some, repudiated by others, *Influ-ences* – and the anthropogeographical scheme it sought to promote – was at turns lauded and criticized. The antithetical responses to Semple's ideas mirrored a differential geography of her book's read-ing. Just as the reaction to *Influences* varied between and within the different contexts of its reception – institutional, disciplinary, and others – so too did its reading. Semple's text was read variously as 'a remarkable book' and 'bunkum' – a volume of 'unquestionable merit' but one also that was verbose and 'a little exhausting.'[10] The dis-similar, sometimes contrary responses that *Influences* provoked are striking and curious. Quite why Semple's book and the ideas it con-veyed meant different things to different readers, and what these differences in reception reveal about the circulation and consumption of environmentalist thought in geography, is this book's central con-cern.

In describing and explaining the history and geography of *Influ-ences'* reception, this book reveals how the ideas Semple sought to communicate were differently understood, staged, and disputed by her different audiences. Bringing together evidence from published reviews, letters, diaries, and marginalia – the material traces of indi-vidual reading experience – I explain why *Influences* was read in cer-tain ways and how these unique interpretations combined within geography's different institutional spaces to constitute shared and common understandings of anthropogeography, whether critical or supportive. My concern is to outline a geography of Semple's book that addresses more than simply the act of reading. In tracing the circulation of anthropogeography – in its textual guise and in its other representational forms – I demonstrate that the reception of Semple's ideas was not delineated straightforwardly by geographical scale (metropolitan, regional, or national, for example), but rather by disciplinary networks defined by shared professional relationships, common visions for geographical research, and friendships forged variously in the field and lecture theatre.

In reflecting upon individual readings within networks of knowledge exchange, I consider how the geography of the book, as a question of reception, might attend both to the spatialized practices of reading and to the spatially-transcendent qualities of interpretative communities.[11] Informed by recent work in the history of science on the reception of scientific texts and by emergent scholarship on the geography of the book, I examine the locational particularities of *Influences'* reception revealed by its incorporation into, and excision from, geography's disciplinary agenda in the United States and Britain. In its geographical attention to the circulation of Semple's text, this book is a study of what one historian of science has called 'knowledge in transit.'[12]

GEOGRAPHIES OF THE BOOK

The historiographical study of the book as a material artefact has changed significantly since the publication in 1958 of Lucien Febvre's and Henri-Jean Martin's agenda-setting *The Coming of the Book*.[13] From its initial focus of the mechanical elements of print – the number and location of printing presses, the manufacture of paper, the distribution and sale of texts – book history has attended increasingly to the social bases of book production and circulation, authorship and reading, textual reception and the exchange of knowledge.[14] Throughout, the work of book history has been underpinned by a spatial sensibility – its attempts to describe and elucidate the social processes which govern authorship, publishing, and reading have been informed by an appreciation of space and situation. Geography, whether manifest in discussions of the location of printing presses, or in analyses of the circulation and consumption of texts, has been central to the historical study of print.

Whilst it has be claimed that 'the geography of the book is still making up its rules', recent studies concerned with print's geographical components have defined something of book geography's potential epistemic and methodological scope.[15] Bertrum MacDonald and Fiona Black have, for example, pioneered spatial analytical techniques in the study of print, employing Geographical Information Science to describe the history of the book in Canada.[16] James Secord and David Livingstone, meanwhile, have set out an intellectual framework for the geography of reading.[17] Still other studies have addressed the geography of (and in) printed texts as a means to understand variously the operation of travel, trade, and empire.[18] Likewise, historians of science have turned to the book to better understand the processes by which scientific knowledge is communicated and received.[19] Bibliographical studies – attending variously to Darwinian evolution, Einsteinian relativity, and Newtonian physics – have exposed national and cultural differences in the reception of science.[20] Work by Nicolaas Rupke on the critical reception of Alexander von Humboldt's *Essai politique sur le royaume de la Nouvelle-Espagne* (1808–1811) in Britain, France, and Germany has, for example, revealed the influence of national reviewing cultures upon the popular conception of Humboldt and his writings.[21] Differences in the apprehension of Humboldt's work have been shown to exist between nations but also between periodicals, reflecting the particular concerns of journals, their authors, and audiences.

In work of this kind on the reception of knowledge, the national has typically served as 'natural unit of assessment' by which the acceptance and repudiation of scientific work is judged.[22] At this scale a tendency exists, however, to homogenize heterogeneous reading practise by assuming national commonalities in the response to books. In an effort to counter this supposition, several studies on the reception of scientific and theological texts have attended in more-or-less explicit ways to local and individual responses to books.[23] Important to this localist turn in reception study is James Secord's *Victorian*

Sensation (2000), which deals with the authorship, publication, and reading of Robert Chambers' anonymously-issued *Vestiges of the Natural History of Creation* (1844). Different patterns of *Vestiges'* reception are shown to exist at different scales of analysis; the meanings attributed to Chambers' book varied 'within regions and between them, within cities and between them, within neighbourhoods and between them.'[24] For phrenologists in Edinburgh, philanthropists in Liverpool, and middle-class women in London, *Vestiges* meant different things; its significance depended upon the particular social, religious, political, and economic contexts within which it was encountered. These distinct engagements reflect 'geographies of reading' and demonstrate the situated nature of reading practice and interpretative communities.[25]

Owen Gingerich's work on the reception of Nicolas Copernicus' *De Revolutionibus* has been similarly attentive to the sites of book production, transmission, and reading.[26] In a wide-ranging census of the approximately six hundred extant copies of the first and second editions of *De Revolutionibus*, Gingerich has interrogated provenance and marginal annotations to describe the invisible college – the intellectual network of students, tutors, and corresponding colleagues – within which Copernicus' ideas circulated.[27] Gingerich makes clear that the reception of *De Revolutionibus* depended not only upon its original printed content, but also upon the ways in which individual copies were variously altered and supplemented with their readers' marginal annotations. Gingerich shows that to speak of the reception of *a* book is problematic; it is necessary to attend, as far as is possible, to the reading of *individual* copies of a book – to marginalia and to matters of provenance.

READING THE RECEPTION OF *INFLUENCES*

To recover the reception of *Influences* depends upon bringing together in combination the 'witness of individual readers' – the disparate but complementary sources in which its historical readings are inscribed.[28] Correspondence, diaries, and marginal notes – together with academic and popular reviews, institutional archives, and the records of Semple's publisher – are the sources from which the contours of *Influences'* reading and the reception of anthropogeography are herein described. Each reader brought to Semple's book a unique set of expectations and assumptions – preconceptions which were conditioned, in part, by the wider intellectual concerns of the social and academic communities to which they belonged. The reading of Semple's book was thus both an individual activity and a collective phenomenon.

How the ideas in *Influences* were received depended upon a reader's individual orientation *and* his or her broader intellectual context. By collating and comparing the disparate indicators of *Influences'* readings, and by treating these individual engagements with Semple's book as part of the wider reception of her ideas, I show that it is possible to make claims about the significance of *Influences* – for particular individuals, for certain academic and intellectual communities, and for specific historical moments – that do not depend upon a national scale of analysis. By highlighting the important role played in the dissemination of scientific ideas by social networks which transcended neat spatial categories, this book questions the appropriateness of scale as an analytical framework for reconstructing the reception of knowledge. My intention is to outline a geographical approach to the study of reception that attends to individual readings of *Influences*, whilst also explaining how and why collective understandings of Semple's anthropogeography emerged. In this respect, this book is not just about reading, but rather it is about the processes – textual and otherwise – which facilitated the reception of Semple's ideas.

Before turning to the reading of *Influences*, I explore in Chapter 2 the intellectual concerns of which Semple's work was a component: environmentalism and anthropogeography. I introduce Semple's biography, describing her educational background and the development of her academic interests, paying particular attention to her work with Ratzel and to her engagement with geography's environmentalist traditions. In an effort to contextualize Semple's work, and this book's reading of its reception, I consider the history of environmentalist thought, and the role of nature-culture theory in shaping the development of disciplinary geography in Europe and North America. Dealing specifically with Semple's attempt to represent Ratzel's anthropogeography to the Anglo-American geographical community, I trace the parallel emergence of disciplinary geography in the United States and United Kingdom, and the development in each of distinct environmentalist research agenda.

Chapter 3 considers the popular and academic reception of *Influences* revealed in published reviews of Semple's book. I examine the relative influence of reviewing cultures – whether defined in terms of national responses to Semple's work, or by assessments which were conditioned by the city, genre, or discipline in which they were composed. Whilst the principal aim here is to discuss and to assess the geography of these reviews, and what they reveal about the initial response to Semple's text in different contexts, questions of authorship, intended audience, and editorial mediation will be shown to be significant in shaping the content of published critiques. The reception of *Influences* was not a matter simply, however, of its reading. It depended to an important extent upon Semple's communication of her ideas in public lectures, scholarly seminars, summer schools, and university lecture rooms. Chapter 4 considers, therefore, the influence of this public oration upon the acceptance (or not) of her anthropogeography, particularly in the half decade immediately following the publication of *Influences*.

Chapter 5 examines the various ways in which Semple's book was employed pedagogically in the United States and United Kingdom. Drawing upon individual reading experiences, these engagements are situated within the educational contexts in which they occurred, and related to then-contemporary geographical debates. My focus is, at turns, biographical and prosopographical: I am interested in both the individual encounters with Semple's text, and in its incorporation into the teaching of geography at different academic institutions. In this respect, this chapter follows not only the trajectory of *Influences'* textbook career, but also describes the locational, institutional, and individual particularities in the discipline's engagement with questions of environmental influence. The book concludes, in Chapter 6, by reflecting upon the role of geography in describing and explaining the differential reception of scientific texts and the situated circulation of knowledge.

2 Anthropogeography: a biography

Ellen Churchill Semple (Figure 2) was born during the American Civil War, in Louisville, Kentucky on 8 January 1863.[1] Her birth closely followed Abraham Lincoln's Emancipation Proclamation – a pair of executive orders that declared free all slaves in territories in rebellion against the federal government (but which did not apply to slaves in border states supporting the Union or to slaves in southern states under Union control). Kentucky was something of a political pivot in this movement, since it represented both the physical and ideological boundary between the Confederate South and Unionist North. Just as territorial control of Kentucky passed between Unionist and Confederate forces at various times during the War, so too did the majority political opinion of its population. Kentucky's unique geographical position conditioned both its role in, and response to, the Civil War. The interrelatedness of geography, politics, and historical events which the War thus revealed would later become for Semple an important research concern and a spur to her interest in environmentalism.[2]

Critical appraisals of geography's historiographical practices have shown that to assume a connection such as this – between Semple's life experiences and her geographical philosophy, for example – reflects a presentist impulse whereby past events are deemed significant only where they are believed to have had some bearing on an important later event.[3] This tendency has been apparent in disciplinary histories which have tended to 'personalize and institutionalize' – to see the development of geographical thought as a retro-

2 Ellen Churchill Semple, 1914

spective question of 'leadership and schools.'[4] An essentialist and potentially problematic implication of such an approach is that intellectual influence is seen to be something which can be attributed straightforwardly to an individual or institution. For this reason, we must treat with caution the idea that the history of anthropogeography as a method and manifesto can be traced simply through the personal histories of Semple and her intellectual mentor, Friedrich Ratzel. This chapter contends however, that the development, articulation, and dissemination of anthropogeography was intimately related to Semple's and Ratzel's intellectual experiences and that it was, perhaps just as importantly, at base a question of texts written and read.

In what follows, I use Semple's biography as the starting point from which to trace the emergence of her geographical interests; to describe how these intersected with those of Ratzel; and to position her anthropogeography in relation to traditions of environmentalist thought. This is neither an essentialist account of Semple's work, nor of geography's engagement with environmentalism, but rather an effort to understand how her geographical interests were initiated, informed, and communicated. In presenting a biographically-informed assessment of anthropogeography, I show that Semple's personal trajectory mattered to the development and articulation of her ideas, and that, as a consequence, there is value beyond the enumerative in engaging with the 'annoyingly complex and uncertain' facets of past geographers', lives.[5] To understand anthropogeography we must understand its progenitors. This chapter is as much, then, about the nature of geographers as it is about the nature of geography.

A BIOGRAPHICAL SKETCH

Semple, the youngest child of Alexander Bonner Semple (1805–1875) and Emerin Price Semple (1822–1904), was spared first-hand experience of the Civil War since Kentucky was, from 1862, controlled exclusively by Unionist forces.[6] Her father, an entrepreneur, operated a business specializing in 'hardware, cutlery, and guns.'[7] Louisville's geographical situation – bordering Illinois on the Ohio River, a major tributary of the Mississippi – facilitated trade with both northern and southern states.[8] As a consequence, Alexander Semple's firm enjoyed considerable commercial success. At Ellen's birth, his family was financially secure: they enjoyed 'good schooling, an abundance of books, and a healthy, well-ordered life.'[9] Although Semple's parents separated during her girlhood, this seems not to have affected the family's financial status – her mother was part of a 'famous Kentucky family' and seems to have drawn upon the support of her extended lineage.[10] Following her parents' separation, Semple passed her childhood within a predominantly female milieu. Within this matriarchy, Semple's mother, 'an exceptionally gifted woman of rare charm', exerted an important and enduring influence.[11] Under her guidance Semple came to 'delight in reading books, especially books on history and travel.'[12] She also mastered tennis and horseback riding, and was, to her apparent reluctance, introduced to Louisville's postbellum social scene.

Her family's wealth and privilege afforded Semple excellent educational opportunities and facilitated her later independent research. She attended a number of girls' schools in Louisville and, to prepare herself for future study, received private tuition. She also 'engaged in a great amount of systematic reading in economics, social science, and history.'[13] In autumn 1878, aged fifteen, Semple followed her sister's lead and entered Vassar College in Poughkeepsie, New York.[14] Although underage, Semple passed the entrance examination – which included questions on geography, grammar, English literature, American history, arithmetic, algebra, geometry, and Latin – 'without conditions' and enrolled as the youngest of three hundred incoming students.[15] Although the college did not offer tuition in geography, something of the intellectual contours of Semple's later concerns were outlined in courses on history and economics, as well as in her training in classic and modern languages. Vassar instilled in her an ability to organize data, to draw conclusions, and to communicate her ideas. The college served also to extend Semple's social circle. Her academic colleagues, many of whom became firm friends, comprised students from a variety of backgrounds, both from the United States and abroad.[16]

Semple graduated from Vassar in 1883 with 'an outstanding record' and was awarded a Bachelor of Arts degree.[17] Her success earned her the position of valedictorian.[18] She delivered a commencement address on 'The conscience of science' at her graduation ceremony. Semple returned to Louisville shortly thereafter and spent a pleasurable but intellectually unsatisfying decade mired in 'a whirl of frequent and elaborate social activities.'[19] After a period of travel in Europe, she found scholarly fulfilment by offering tuition at her sister's private school in Louisville – the Semple Collegiate School.[20] Despite Semple's rather intimidating persona – 'slim and straight, with masses of dark hair, the crispest white collar and jabot' and 'an air of almost fierce authority' – she was regarded fondly by her students, who knew her as Miss Nelly.[21] She taught Latin, ancient history, and physical geography, and seemed to revel in the task. As one student recalled,

> She really enjoyed teaching. She loved to see her class catch fire. The pains that she used to take with the stupid as well as the brilliant paid off for she held her group of girls enthralled, at times even frightened by her zeal for imparting knowledge. And what knowledge![22]

Semple's pedagogic ability, honed in preparing Louisville's privileged for college entrance examinations, would later be crucial to the propagation of her geographical philosophy.

Whilst the experience of tutoring proved valuable to Semple's later work, it was not her career's 'motivating germ.'[23] Inspiration came in this regard from discussions with two 'widely read and cultivated lawyers and a brilliant Jewish Rabbi.'[24] Conversations with these men, supplemented by access to their libraries, had an important influence on Semple's intellectual development.[25] She became interested in questions of environmental influence, but found little beyond the 'purely pseudo-scientific writings of Henry Buckle' with which to engage.[26] Buckle's work, expounded in his unfinished *History of Civilization in England* (1857–1861), posited the view that man, to use Buckle's terminology, progressed from 'a stage in which he was completely dominated by the environment to one in which he had obtained freedom from it and even controlled it.'[27] The ultimate stage in this progression was represented, in Buckle's model, by mid-nineteenth-century Western Europe. Whilst Buckle's work appealed to Semple in terms of the answers it seemed to provide on the subject of environmental influence, she was 'shrewd enough to see ... that he was not authoritative. He was valuable only as being suggestive.'[28] As she later put it, 'I began to scent the importance of geographic influences, tho' at that time ... I struck no trail of a previous investigator that was reliable enough to follow.'[29]

In 1887 Semple again visited Europe – travelling to London in the company of her mother. There she was introduced to Duren James Henderson Ward (1851–1942), a recent Ph.D. graduate from the University of Leipzig.[30] Ward relayed news of a charismatic professor of anthropogeography, Friedrich Ratzel (Figure 3), whose lectures 'made history come alive.'[31] In what was later termed the 'turning point in her career', Semple borrowed from Ward a copy of the first volume of Ratzel's *Anthropogeographie* (1882).[32] She kept the book for six months and, supplemented by a bibliography from Ward, systematically absorbed Ratzel's oeuvre. Semple found in Ratzel a compelling approach to geography and history. She resolved to go 'immediately to Leipzig to study under him.'[33] To prepare for research under Ratzel, Semple began external studies towards a Master of Arts degree in Economics and Social Science from Vassar. Throughout this period of supplementary education she maintained a correspondence with Ward, discussing 'many of the then on-going problems of evolution and the influence of various types of environment.'[34] In 1891, having completed a thesis on slavery, Semple travelled to Leipzig to undertake work with Ratzel.

Although Semple had studied German for six years in the United States, she spent her first three months in Leipzig in lodgings with a local family in order to 'get a command of the vernacular.'[35] This linguistic and conversational preparation was important since her admission to study at Leipzig would depend upon her skills of negotiation and persuasion. In the late nineteenth century, the University of Leipzig did not permit female students to matriculate or officially to sit examinations. Female students were, however, permitted to attend lectures and seminar series if they petitioned the organizing faculty directly. A testimonial as to their abilities would be provided upon completion of the course.[36] Following a personal application to Ratzel, Semple was afforded *ordentliches Mitglied* (regular member) status, and was admitted to his geographical seminary.[37] She also undertook economic studies with Wilhelm Georg Friedrich Roscher

3 Friedrich Ratzel, undated

(1817–1894), from whom she learned a 'wonderful method in inductive research.'[38] She later credited Roscher for ensuring the rigour of her work: 'Roscher taught me to take off my hat to every obstructive fact that threatened to block my theory; and that is a great debt to owe any man.'[39] It was from Ratzel, 'my inspiration, my dear master and friend', that she drew personal and professional motivation.[40]

Whilst the typical Leipzig professor was dour and efficient, Ratzel was an enthusiastic and effective orator. For Joseph Russell Smith (1874–1966), an American geographer who studied at Leipzig a decade after Semple, Ratzel was 'as full of energy as a steam engine. He bounces along like a boy.'[41] Something of the influence of Ratzel's lectures came, then, from his convincing and charismatic presentation, rather than from the straightforward communication of his geographical principles. His oratorical skill was apparent when, in 1902, Wilhelm II visited the university and chose to attend one of Ratzel's lectures. For one observer, Ratzel was 'by far the most imposing figure in the party, he was not only the most handsome, but he was the most learned and altogether the most kingly.'[42]

Whilst Ratzel's enthusiasm and apparent regal confidence enthralled Semple as much as it did his other students, she was attracted more particularly to his geographical approach. Although a number of biographical treatments have asserted that as a consequence of her sex Semple was required to sit 'in an adjoining room, with the communicating door ajar' when attending Ratzel's lectures, this was not the case.[43] The pair enjoyed a constructive and collegiate relationship – discussing, among much else, 'the philosophy of style, and style in geographical writing.'[44] Such discussions were critical to Semple's later articulation of anthropogeography. As she recalled,

> Ratzel, in his frequent talks with me, urged the value of a literary style for books on Anthropo-geography. He argued that since the science had to do with man, it was entitled to the same literary treatment as History. I took his admonitions to heart, not only because I agreed with him in theory, but also because I anticipated that anthropo-geography would make its way slowly in this country [the United States], and that outward charm might help to secure for it more open doors.[45]

In contemplating Semple's response to Ratzel's geographical work, and her desire to communicate it to an Anglophone audience, I turn now to consider the development of his philosophy in the context of nineteenth-century German geography.

RATZEL AND THE DEVELOPMENT OF ANTHROPOGEOGRAPHY

The contours of German geographical investigation in the eighteenth and nineteenth centuries were drawn largely by Alexander von Humboldt (1769–1859) and Carl Ritter (1779–1859).[46] Building upon the discipline's philosophical foundations – which had earlier been laid by the Prussian geographer Immanuel Kant (1724–1804) – Humboldt and Ritter attempted to extend its scholarly purview by outlining novel methodological approaches to the study of geographical features. Humboldt's exploration of tropical America at the turn of the nineteenth century served to codify a systematic and instrumental approach to natural science and to 'further systematize the theory of the control of land-forms and climate over the distribution and habits of plants, animals, and man.'[47] Humboldt's contribution was, then, to questions of environmentalism.[48] The issue of how far and in what ways the earth's physical features influence humankind's social and physical development is part of a long debate within Western intellectual traditions, traceable to Hippocrates and Aristotle.[49] From the beginning of the nineteenth century, however, it became (principally as a consequence of Humboldt's explorations) a question *of* geography and a question *for* geography.[50]

In contrast to the peripatetic Humboldt, Carl Ritter, an 'armchair geographer', spent the majority of his professional career at the University of Berlin, where in 1820 he was appointed to the first chair of geography.[51] Ritter's principal influence upon the development of geography as an academic discipline was in his criticism of the descriptive nature of geographical investigation. He advocated a scientific and inductive approach to geography which he termed *Erdkunde* (earth science).[52] Ritter's ideas were most fully expressed in his unfinished nineteen-volume *Die Erdkunde im Verhältniss zur Natur und zur Geschichte des Menschen* (The science of the earth in relation to nature and the history of mankind), written between 1817 and 1859.[53] Infused with a teleological vision, Ritter's work was an attempt to apply a physiological approach to the study of the earth, in order that the laws which govern it might be discerned. The metaphor of comparative anatomy was apparent in his analysis and classification of regional difference. For Ritter, it was in the comparison of similar regions in different parts of the world – a form of areal differentiation

– that causal relationships describing the 'reciprocal and evolutionary relation of environment and society' could be apprehended.[54] Ritter's geography conceived of the earth's surface as a series of 'discrete and objective natural regions' which were 'uniquely linked to a particular national ethos.'[55] Ritter's contribution to geography was principally two-fold: he outlined a systematic and inductive methodology and, in 'conjoining Land and Volk', facilitated a geographical understanding which took land and its inhabitants to be intimately correlated.[56]

Ritter and Humboldt both died in Berlin in 1859, the year of Charles Darwin's *On the Origin of Species*.[57] Their deaths marked the end of a period of significant development in German geography and heralded 'the beginning of a crisis in scientific and philosophic thought.'[58] Ritter's chair of geography at Berlin remained vacant until 1874, and geography became 'a side issue in the curricula.'[59] Something of a disciplinary focus was recovered in 1871 when – following the Franco-German war and the establishment of a German state – a chair in geography was created at the University of Leipzig.[60] Its first occupant was Oscar Peschel (1826–1875). In contrast to the Kantian idealism which permeated the work of Humboldt and Ritter (the notion that our knowledge of objects is inherently subjective), Peschel was influenced by the philosophical doctrine of materialism, which holds that what is observed in nature can be explained only by reference to natural causes, not by assuming the existence of an external, supernatural power. Peschel was critical, therefore, of the apparent teleological basis to Ritter's work and sought to offer a materialist revision to his *vergleichende Erdkunde* (comparative geography). Peschel's principal contribution was to revise, and to define more narrowly, the basic units of comparative analysis. Where Ritter sought comparisons at a continental scale, Peschel attended to particular types of landforms – valleys, mountains, glaciers, lakes, fjords, and so on. In this respect, Peschel's morphological focus was an important prompt to the development of systematic physical geography.

Aspects of Peschel's research focus were developed by a former student of Ritter – Ferdinand von Richthofen (1833–1905).[61] A geologist by training, Richthofen undertook the majority of his field research in Southeast Asia, the details of which occupied his five-volume *China: Ergebnisse eigener Reisen und darauf gegründeter Studien*

(China: the results of my own travels and studies based thereon) published between 1877 and 1912. Richthofen made a special study of loess (fine-grained, wind-blown soil) as both a geological phenomenon and as evidence of the 'reciprocal action of man and his environment.'[62] Following his research in China, Richthofen pursued an academic career. He was appointed Professor of Geology at Bonn in 1875, before succeeding Peschel at Leipzig in 1883. Richthofen used the opportunity of his inaugural lecture at Leipzig to communicate his manifesto for the scope and method of geography. His geography was based upon the intensive observation and description of the earth's surface features, and an attempt to relate these to their physical underpinnings – that is, for example, to describe the development of soil by reference to the base geology. In this way, Richthofen's scheme allowed for both descriptive geography (special geography or chorography) and explanatory geography (general geography or chorology). He saw geography as both idiographic (concerned with the description of unique features) and nomothetic (concerned with generalities and the laws governing them).

The systematic study of the earth's physical features pioneered by Peschel and Richthofen informed Ratzel's approach to human geography. In much the same way that Peschel and Richthofen advocated the comparative study of representative landforms, so too was Ratzel concerned to understand the 'biotic and cultural' features of different social groups in relation to their environment.[63] Ratzel's intention was to do for human geography what Peschel and Richthofen had done for physical geography – that is, to make it a science. His approach to human geography was informed, to a significant extent, by his student training in natural history and zoology.[64]

Ratzel began his academic training in the mid 1860s – variously at the universities in Heidelberg, Jena, and Berlin – approximately five years after Darwin's *On the Origin of Species* had 'newly invigorated' the natural sciences.[65] His enthusiasm for the Darwinian method – particularly as interpreted by Ernst Heinrich Haeckel (1834–1919), his graduate professor of biology at the University of Jena – was expressed in his first book *Sein und Werden der organischen Welt. Eine populäre Schöpfungsgeschichte* (The nature and development of the organic world. A popular history of creation) published in 1869.[66] Haeckel's interpretation of Darwin facilitated a world view in which all organic life could be explained and understood by reference to the 'natural laws and processes that had been proclaimed by Darwin.'[67] In this respect, human beings could be subject to study in much the same way as other animal or vegetable life – the fundamental controls on their development being the same. Darwin's work, as interpreted by Haeckel, provided a basis by which human adaptation to the environment might be described and explained scientifically as a question of biology.

Ratzel's enthusiasm for the Darwinian approach was further re-inforced during a short period of study at the University of Munich, where he was introduced to the naturalist and ethnographer Moritz Wagner (1813–1887).[68] Based upon fieldwork in Central America during the late 1850s, Wagner had formulated a theory to describe the function of the migration of species in the development of organic diversity.[69] Wagner was somewhat critical of the principles of natural selection and proposed, instead, a *Migrationstheorie* (law of migration) which stated that it was the dispersal of organisms across space, and into new environments, that facilitated adaptation, and that these adaptations were subsequently preserved by means of geographical isolation.[70] Wagner's perspective gave Ratzel 'his first direct awareness of the interest of geographical work.'[71] In the years immediately following his exposure to Wagner's theory, Ratzel undertook a number of important expeditions as travel correspondent for the *Kölnische Zeitung* (Cologne Journal). The most significant of these was his 1874–75 sojourn in North America.[72] Of principal interest to the readers of Ratzel's dispatches in the *Kölnische Zeitung* was the contribution of German migrants to the United States, particularly their role in the nation's westward expansion. Ratzel made a special study of other minority groups, including the settlement of the Pacific coast states by East Asians. This period of field observation was critical for the later development of his anthropogeographical principles; it helped to clarify his understanding of 'the relationship between the political state and its environmental milieu.'[73]

The publication of *Anthropogeographie*

Following his appointment to a lectureship in geography at the Technische Hochschule (Institute of Technology) in Munich in 1875, Ratzel began to formulate his perspective on human geography (and, in so doing, to make a substantive and systematic contribution to the discipline).[74] The principal aim of Ratzel's developing geographical philosophy was to refute Buckle's claim that 'as civilization advances it becomes more & more divorced from the physical environment.'[75] Ratzel saw modern civilization as 'a product of the close interrelationship between culture & environment' – not as a separate and independent phenomenon.[76] He was influenced in his thinking by the social evolutionary writings of Herbert Spencer (1820–1903). Spencer's perspective on human societal development – expounded most fully in his *First Principles* (1862) – was based upon biological evolutionary principles.[77] For Spencer, the development of human society (and, by implication, the state) was analogous to organic evolution in that competition for survival and predominance facilitated and prompted adaptation. Aspects of this organic conception were apparent in Ratzel's numerous publications on the United States, but expressed most systematically in *Anthropogeographie, oder Grundzüge der Anwendung der Erdkunde auf die Geschichte* (Anthropogeography, or the principles of the application of geography to history) (1882).

In the first volume of *Anthropogeographie*, Ratzel sought to describe how the distribution and comparative success of human populations could be seen as 'more or less' a function of environmental conditions.[78] In this respect, Ratzel's text was a reworking of Ritter's *Erdkunde*, and an attempt to elevate the study of human-environment relations above the pseudoscience of Buckle.[79] Ratzel was keen to show that as societies developed they 'became more and more enmeshed in their lands.'[80] By eliminating the teleological framework associated with Ritter's geography, Ratzel presented a model of human development that was, in effect, directionless and without 'ultimate purpose.'[81] Since there was no divine direction, the evolution of human societies was seen to depend upon a Spencerian struggle for survival, with those most adept at responding to the challenges and opportunities afforded by particular environmental circumstances being ultimately successful. In this respect, Ratzel's scheme was coloured by a 'deterministic tint.'[82]

Combining Wagner's *Migrationsgesetz* with the biological evolutionary ideas of Spencer and Jean-Baptiste Lamarck (1744–1829) – or, more properly, their application to the understanding of the social organism – Ratzel's volume was a synthesis of the theoretical positions underpinning his perspective and an attempt to set out systematically what he took the study of *Anthropogeographie* to be.[83] His intended foci were threefold: the distribution of human societies on the earth's surface; the function of migration and the environment in relation to these distributions; and the developmental influence of the physical environment upon individuals and social groups. Aware that 'the geographer cannot formulate laws expressed with mathematical precision', Ratzel's work was an attempt to describe how the study of human geography might proceed rather than a proof of his concept.[84] As a methodological statement Ratzel's text was received as a timely conformation 'of the view then held by scientists, that environment determined the characteristics and the line of development of a people.'[85] Yet, the validity and value of Ratzel's geography when applied to research in the field was, for some readers, uncertain. This much was true of the German anthropologist Franz Boas (1858–1942).

Despite having received a doctorate in physics from the University of Kiel in 1881, Boas 'self-identified as a geographer.'[86] His education was richly infused with geographical themes. As a doctoral student, for example, he produced work on 'the northern limit of Greenland, and geography as the necessary foundation of history.'[87] At Kiel, and previously at the University of Bonn, Boas had come under the influence of Theobald Fischer (1846–1910), a disciple of Ritter who lectured on geographical exploration and polar research. Fischer's influence – combined with Boas' interest in questions of environmental influence – led him to undertake an expedition to Baffin Land (now Baffin Island) with the purpose of confirming the environmentalist position then current in German geography. Ratzel's *Anthropogeographie* provided Boas with a 'systematic representation of the ideas which I had then in mind, and which I desired to study in one particular field.'[88] His investigation of the social organization of Baffin Land's Inuit (or Eskimo as they were then called) showed that 'in the same physical environment different cultural forms occur.'[89] He concluded that 'the environment can only act upon a specific culture, not determine it.'[90] Something of this perspective was rehearsed in his 1887 methodological treatise 'The study of geography', published in *Science*, in which Boas discussed the segregation of geography between descriptive and nomothetic research.[91]

Ratzel succeeded Richthofen at Leipzig in 1886, and in lectures to small seminar groups perfected his pedagogic style. At Leipzig, Ratzel also composed the second volume of his *Anthropogeographie*, subtitled *Die geographische Verbreitung des Menschen* (the geographical distribution of mankind), which was published in 1891. Ratzel's previous proclamations on anthropogeography had been subject to criticism by Hermann Wagner (1840–1929), professor of geography at the University of Göttingen, principally because 'the basis in data and the citation of authorities was too slender.'[92] Wagner, who corresponded with Ratzel, worked closely also with Franz Boas on the *Geographisches Jahrbuch* – an annual geographical bibliography.[93] It seems probable that Ratzel was aware of the concerns expressed as to the empirical validity of his work, and sought to provide a 'foundation in fact.'[94] The second volume of *Anthropogeographie* was, then, situated more firmly in data – indeed Ratzel devoted an entire section of the book to a discussion of population statistics.[95] It dispensed, moreover, with the deterministic environmentalism which characterized the preceding volume and was rather more constrained in terms of its theoretical pronouncements. Where the first volume dealt principally with the effect of the physical environment upon human history, the second attended more particularly to the social organisation of human societies in relation to their environment. This subtle distinction was important, since it later facilitated both deterministic and possibilistic interpretations of Ratzel's ideas.[96] Ratzel later issued a second edition of the first volume of *Anthropogeographie*, which brought it into closer alignment with the second volume.[97]

Semple's period of study at Leipzig between 1891 and 1892 coincided with the ultimate expression of Ratzel's anthropogeography. His four thematic principles – that human societies develop 'within a frame (*Rahmen*), exploiting a place (*Stelle*), needing space (*Raum*) and finding limits (*Grenzen*)' – became the fundamental tenets of Semple's later discussions on anthropogeography.[98] Whilst in Ratzel's lectures,

Semple 'avidly absorbed whatever the master propounded and truly became his disciple.'[99] As her appreciation for Ratzel's anthropo-geography developed, Semple became concerned to communicate it to 'an American public to whom the subject was quite new.'[100] She was conscious that since there was 'no previous Ritter and Peschel on this side of the water [the United States]', she would be required first to outline the basic principles of a systematic human geography. Moreover, given that Ratzel's work was 'so closely adapted to conditions obtaining in Teutonic and Slavonic Europe', it was 'to most American and English students ... a closed book.'[101] Semple's project was, then, one of translation, clarification, and adaptation – of bringing Ratzel's work to the attention of a different interpretative community.

BRINGING ANTHROPOGEOGRAPHY TO THE UNITED STATES

In the eight years since Semple had been introduced to Ratzel's work by Duren Ward, she had evolved from pupil to preceptor: 'Previously she had followed; now she would lead the anthropo-geographic movement in the United States.'[102] Semple's central project upon returning to Louisville in the mid 1890s was to communicate Ratzel's ideas to the English-speaking world, 'but clarified and reorganized.'[103] To that end she devoted her time to library study and field research, honing 'the craft of authorship.'[104] To further her authorial skills, Semple joined the Authors' Club of Louisville – a recently-formed cabal of aspiring female writers who composed fiction inspired by Louisville and its environs.[105] Semple was, in this way, able to refine the literary style which both she and Ratzel believed necessary to communicate the anthropogeographical position.

Semple's initial academic writing was confined to the direct translation of German scholars – principally the economist Karl Diehl (1864–1943) and the sociologist Ludwig Gumplowicz (1838–1909), both of whom published a number of reviews in the *Annals of the American Academy of Political and Social Science*. Semple also translated short works by Ratzel and published reviews of his work.[106] It was in the lecture theatre, however, that Semple was first able to synthesize and represent aspects of Ratzel's geographical philosophy. In common with her desire to reframe Ratzel's arguments in a locally-tailored form, Semple presented a lecture entitled 'Civilization is at bottom an economic fact' at the Third Biennial General Federation of Women's Clubs in Louisville on 29 May 1896. Her talk – said to be 'one of the most valuable papers of the convention' – was delivered under the auspices of the Philanthropy and Home section of the Federation, and was intended to address questions of economic disparity that were then apparent in Louisville.[107]

Semple's first opportunity to address a more obviously geographical audience came in 1897 with a paper she contributed to the first volume of the *Journal of School Geography*. The *Journal* had been established that year by Richard Elwood Dodge (1868–1952), a former student of William Morris Davis. Dodge was then professor of geography at Teachers College, Columbia University, and along with Davis was part of 'that general movement ... which created modern geography in the United States.'[108] Semple's paper, 'The influence of the Appalachian barrier upon colonial history', was an attempt to apply something of Ratzel's method to the historical study of North America.[109] The Appalachian barrier – running from Vermont to Alabama – represented, for Semple, an impediment to the historical settlement of the continent and an environment within which early colonial settlers were 'protected from without by bulwarks of nature's own making.'[110] In Semple's formulation the geographical arrangement of the continent was seen to confer upon British settlers in the Thirteen Colonies a 'certain solidarity which they would not have otherwise possessed' – a factor vital, she believed, in their ultimate success in the American Revolutionary War.

The environmentalism of Turner and Shaler

Semple's use of environmentalist principles to explain the historical development of the United States was not, however, unprecedented. At a meeting of the American Historical Association in Chicago in 1893 – organized in conjunction with the World's Columbian Exposition, celebrating the four-hundredth anniversary of Columbus' voyage – Frederick Jackson Turner (1861–1932) had presented a 'penetrating essay' on 'The significance of the frontier in American history'.[111] Turner's paper, it is claimed, 'became the most famous scholarly paper ever delivered by an American historian.'[112] More prosaically, it motivated an intellectual reassessment of the nation's frontier experience. Turner's thesis saw the American West as a metaphor and an explanation for the distinctive historical development of the United States. For Turner, the 'ever retreating frontier of free land' to the West of the Appalachians was fundamental to the nation's historical experience and social development.[113] The physical and cultural distance which separated the frontier from the eastern seats of power promoted an individualism and ad hoc democracy among the frontier's pioneers.[114] Reliance was placed upon individual wit and strength, and centralized political control was regarded with suspicion. In this respect, the frontier was seen to be responsible for facilitating a national character – and consequently national institutions – in which individual liberty was emphasized.[115] In Turner's scheme, 'bio-social inheritance was envisaged as subservient to the influence of the physical environment in shaping the American nation.'[116]

Turner's personal background was not dissimilar to Semple's. He passed his boyhood during the Civil War in Portage, Wisconsin. Like Louisville, Portage represented a 'semifrontier milieu' and was an important centre of commerce.[117] Turner's hometown, as Louisville had done for Semple, provided 'a typical example of the theory of American history to which he devoted his life.'[118] Turner was educated in zoology, botany, physics, and chemistry at the University of Wisconsin, and received his doctorate from Johns Hopkins University.[119] His principal intellectual influences were evolutionary and environmentalist: he drew variously upon Darwin, Spencer, and Thomas Henry Huxley (1825–1895).[120] A particular spur to the development of his frontier thesis was, however, the deterministic political economy of Achille Loria (1857–1943), an Italian economist whom Turner read whilst a young professor of history at Wisconsin (1890–1910).[121] Loria, who was also influenced by a Spencerian perspective, saw economic development as a function of the relative scarcity or abundance of land.[122] In this respect, Turner's formulation, and Loria's too, was similar to the anthropogeographical principles outlined by Ratzel.[123] In applying to the historical study of the United States the Lam-

arckian metaphor of the social organism, Turner's perspective corresponded with the 'new science of evolutionary human geography.'[124] As such, Turner (who was admitted to the Association of
American Geographers as a full member in 1915) has been credited as
the 'cofounder, along with Ellen Churchill Semple and Albert Perry
Brigham' of the subfield of American geography concerned with
environmental influence.[125] This is, though, a historiographical con-
ceit: although Brigham, Turner, and Semple understood their work to
be complementary, they did not consider the establishment of an
environmentalist geography to be their common aim.

Whilst Turner's and Semple's contributions to the post-Darwinian
project can be seen to have exerted a novel influence upon the disciplinary focus of American geography, aspects of their intellectual
interests were apparent in the earlier 'creative outlooks' of George
Perkins Marsh (1801–1882) and Nathaniel Southgate Shaler (1841–
1906).[126] Turner's work can be seen to have built, most particularly,
upon that of Shaler, a Kentucky geologist-geographer who studied
under the Swiss naturalist Louis Agassiz (1807–1873) at Harvard and
who later taught William Morris Davis.[127] Among Shaler's most
notable contributions to the environmentalist canon was *Nature and
Man in America*.[128] Shaler's book, situated firmly within the context of
contemporary anthropological science, considered the relationship
between human society and the physical environment in North
America (particularly the United States). As he noted in the volume's
introduction, 'In the light of modern science, we regard our species as
the product of terrestrial conditions.'[129] Shaler regarded the influence
of the environment upon organic life as a function of its developmental stage – the more advanced the organism, the greater its
dependence upon the environment: 'When the human state is attained ... the relations of life to the geography and other conditions
of environment increase in a wonderfully rapid way.'[130] Formulated
in this way, Shaler's perspective on environmental influence was
'fully consonant' with that of Ratzel – societal development was seen
to be paralleled by increasing dependence upon the physical environment.[131] The thrust of Shaler's book contrasted the 'unsuitability of
the North American continent as a cradle for civilization' with its
suitability as an arena in which incoming races could prosper.[132] In
this way, and echoing Ratzel's earlier work on German immigrants in

the United States, Shaler argued that 'in its transplantation from Europe to America the Aryan race had not deteriorated, but had probably benefited.'[133]

Shaler's engagement with neo-Lamarckian conceptions of social development extended into the analysis of 'mental, moral, and social realms.'[134] This was apparent, most particularly, in his discussion of race and racial superiority. Shaler's views were predicated upon the notion that different racial groups did not share a common origin (or that a common origin was so temporally distant that sufficient modification had taken place in the interim to render races morally and intellectually distinct). This polygenist perspective was underpinned by a number of environmentalist principles, particularly those related to climatic influence and geographical isolation.[135] As he made clear, for example, the emergence of civilization in Europe depended upon 'the stress of the high latitudes, [and] the moral and physical tonic effect of cold.'[136]

Aspects of Shaler's argument had earlier been set out by Marsh in his treatise on the relationship between nature and society: *Man and Nature; or, Physical Geography as Modified by Human Action* (1864).[137] Where Marsh's perspective differed from that of Shaler, however, was in its predominant attention to the deleterious effect of society on the environment.[138] Marsh – later described as the prophet of conservation – developed his geographical perspective having witnessed the destructive effects of deforestation in his home state of Vermont and in Turkey and Italy where he served as a diplomat.[139] Marsh's perspective was articulated, moreover, in opposition to his near-contemporary, the Swiss scholar Arnold Henry Guyot (1807–1884).[140] Guyot had relocated from Europe to the United States in the 1840s, initially delivering a series of lectures at the Lowell Institute in Boston, before accepting a permanent position at Princeton University.[141] Guyot's geographical perspective – articulated in his *Earth and Man: Lectures on Comparative Physical Geography in its Relation to the History of Mankind* (1849) – brought a Ritterian and deterministic model of comparative geography to the attention of the American academy.[142] Whilst Guyot's belief that 'the earth made man' was then uncontroversial, Marsh sought to prove the opposite – that 'man in fact made the earth.'[143] Despite this fundamental difference, Guyot 'applauded Marsh's insights', and his work was commended by Shaler and Davis.[144] In this respect, Guyot and Marsh both exerted an important, although different, influence on geographical thought in the United States. It is apparent, then, that Semple's work on the Appalachian barrier, and Turner's frontier thesis, emerged from, and were representative of, a broader trend within American scholarship (in no small part a corollary of Humboldt's influence on American intellecttual life) which encompassed geography, geology, history, culture, race, politics, and economics.[145] The contributions of Semple and Turner were, then, new rather than entirely original.

Anthropogeography in the field:
investigating 'moonshine whiskey and wretchedly cooked food'

Between 1897 and 1900 Semple contributed five further papers on various aspects of anthropogeography to the *Journal of School Geography*, and one on that subject to the *Journal of the American Geographical Society of New York*.[146] These papers were based largely upon secondary sources and were restatements of ideas she had absorbed during her time in Leipzig rather than original pronouncements. Semple's contributions to the *Journal of School Geography* were considered by its editor, Richard Dodge, to be important in communicating to 'the common school teachers' the principles of anthropogeography.[147] In the journal's view, geography was conceived of as 'the science of man's relation to his earth environment', and the object of its inquiry 'the mutual dependence of man and nature upon one another.'[148] Given this formulation, Semple's anthropogeography could be seen to parallel the initial contours of academic geography in the United States and also to represent a model for the teaching of school geography. Whilst Semple received the support and encouragement of Dodge and Davis in her promotion of Ratzel's geography, she retained a desire to demonstrate the utility of anthropogeography in the field. Semple's perspective on geographical research, 'based on bold and keen creative insights' seemed to offer a model for a systematic engagement with human geography – one which might most convincingly be demonstrated in the field.[149]

In the summer of 1898, Semple participated in a philanthropic project of the Kentucky Federation of Women's Clubs to establish a 'social settlement' for the intended benefit of the nearby populace in Hazard, 'a squalid, wretched little town in the heart of the Kentucky mountains.'[150] Whilst the project was motivated largely by religious and moral imperatives, Semple saw the difficulties of life in the mountains as a problem of geography rather than theology. In the scattered and isolated population, Semple saw clearly the deleterious influence of the physical environment on what was, fundamentally, an Anglo-Saxon population: 'isolated by mountain ranges from the outside world and *from each other*, their naturally fine stock deteriorating constantly from the effect of too close intermarriage, moonshine whiskey and wretchedly cooked food, these people have degenerated in many respects.'[151] Yet, for Semple, the perceived superior lineage of the population ensured that 'in talking to them, one is deeply impressed with the fact that the material is sound and good.'[152] The social settlement took the form of a tent – 'decorated with flags, Japanese lanterns, and photographs of the best pictures' – where books, newspapers, and periodicals were made available, and the basic principles of hygiene and domestic economy communicated.[153] Over the course of its six-week operation, the settlement was judged to have been successful in bringing to the mountains certain aspects of lowland social and religious culture.[154]

A fundamental tenet to which the volunteers in this philanthropic project subscribed was that the Kentucky mountaineers were 'our brothers in blood.'[155] This perspective, inspired by Scripture, depended also upon their shared Anglo-Saxon ancestry. The task of the Federation of Women's Clubs was, then, not to impose an entirely novel social framework on the mountaineers, but to reinvigorate one which had been lost as a consequence of a century's isolation (and also to counter what they saw as the insidious spread of Mormonism and Roman Catholicism in the mountains). This is not to suggest, however, that the work of the Federation was considered a question of racial superiority. The dominant theological model in the Commonwealth of Kentucky during this period was that 'God has made of one blood all peoples of the earth' – a firmly monogenist perspective.[156] For Semple, the notion of an immutable and geographically independent racial superiority (at least in the face of the challenges of the physical environment) made little sense. Her interest in the Kentucky mountaineers was due not so much to the fact that their origin was broadly Anglo-Saxon, but rather that they provided an instructive case study: in little more than a century the 'naturally fine stock' from which they were descended had been so significantly modified as a consequence of geographical isolation that their social, economic, and agricultural systems were largely novel.[157]

Semple's work in the Kentucky Mountains continued in 1899, when she completed a 350-mile horseback journey through its more isolated stretches. Her observations formed the basis of her most personal contribution to the anthropogeographical literature: 'The Anglo-Saxons of the Kentucky Mountains: a study in anthropogeography'.[158] This paper, published in the Royal Geographical Society's *The Geographical Journal*, was said to take 'high rank among the geographical articles in the English language.'[159] Semple's paper introduced to its readers a region where the population was 'still living the frontier life of the backwoods', where Elizabethan English was spoken, and where 'the large majority of inhabitants have never seen a steamboat or a railroad.'[160] The principal characteristic of life in the mountains was a geographical isolation – not only from the rest of the state, but from one another – that left the population 'almost as rooted as trees.'[161] The consequence of this comparative immobility was close intermarriage and the preservation of 'the purest Anglo-Saxon stock in the United States.'[162] The same isolation that facilitated this conservation of racial qualities was seen also to have facilitated a 'retarded civilization' where the 'degenerate symptoms of an arrested development' were apparent.[163]

The physiological effects of the mountain environment were particularly apparent to Semple. The population was seen to have lost the 'ruddy, vigorous appearance' of their forebears, having become 'tall and lanky ... with thin bony faces, sallow skins, and dull hair.'[164] Despite these outward adaptations to the rigorous mountain life, Semple confidently detected 'the inextinguishable excellence of the Anglo-Saxon race.'[165] In this respect, whilst the mountain environment was a limiting factor in the physical and societal development of its population, it did not circumvent entirely the civilizing potential of their genetic inheritance (wrought over millennia amid more favourable conditions). Whilst the physiological effects of the environment were pronounced, Semple saw the influence of the mountain topography most particularly in the vernacular architecture. The most remote and isolated communities typically displayed cabins that were 'primitive in the extreme' and suggestive of 'pioneer archi- tecture.'[166] The barns which accompanied these were, in an echo of their topographical setting, redolent of the 'Alpine dwellings of Switz- erland and Bavaria.'[167] In broader valleys, where access to the centres of sawmilling was easier, the buildings were akin to 'village dwellings in Norway.'[168] For Semple, then, there was an evident parallel be- tween environmental conditions and architectural style in terms both of what could be constructed and what was most suited to local cir- cumstances. The existence of similar dwellings in the Alpine and Ken- tucky Mountains did not necessarily indicate a shared culture, there- fore, but a set of common geographical circumstances.

In physiology, architecture, and social organization, Semple saw the influence of the mountain environment written upon the Ken- tucky highlanders. Geographical isolation and topographical obs- tacles were presented – with implicit reference to the tenets of an- thropogeography – as an explanatory framework: the environment being the basis by which the peculiarities of the mountaineers' soci- ety might be understood and accounted for. Semple's demonstration that anthropogeography could be studied in the field, and that envir- onmental influence was thus an apparently legitimate and demon- strable causal explanation, was significant for those geographers – particularly Davis and Dodge – who believed that the promotion of the discipline depended upon an ability to adhere to a scientific and nomothetic approach.[169] Davis, for one, thought Semple's paper ought to 'serve as the type of many more.'[170] In appearing thus to satisfy Davis' desire for rational and deductive geographical research, Semple's paper drew positive attention: it 'fired more American stu- dents to interest in geography than any other article ever written.'[171] It served also to secure Semple a more prominent position within the professional geographical community. The fact that her work had been published by the Royal Geographical Society brought her to the attention of an international constituency.

FROM 'UNASSUMING LITTLE WOMAN'
TO PROFESSIONAL GEOGRAPHER

Concurrent with her fieldwork in the Kentucky Mountains, Semple was encouraged by Richard Dodge to collate her earlier articles on North America in a single volume.[172] Semple was contracted by the Boston publisher Houghton, Mifflin and Company to produce a book setting out her perspective on the influence of the physical environment upon the course of American history.[173] In preparing the volume, Semple undertook extensive secondary research, visiting 'Washington and ... the magnificent Mercantile Library of St. Louis.'[174] The most valuable and instructive material for her study was found in Louisville, at the private library of Reuben Thomas Durrett (1824–1913). A cofounder of Louisville's Filson Historical Society, Durrett amassed an unparalleled collection of primary material relating to the historical settlement of Kentucky and the Ohio River Valley, and an impressive anthology of secondary literature dealing with travel and historical accounts. Semple sought, by reference to these authorities, to present a convincing demonstration of the application of anthropogeographical principles to the study of the United States. Semple's book attended to the environmental factors which she understood to have conditioned, among other things, war, migration, commercial development, the location of cities, the provision of transportation, and international trade. In tandem with her examination of the Kentucky mountaineers, Semple had demonstrated that an anthropogeographical approach might be applied with equal success to the study of historical and contemporary society.

Semple's original plan had been to entitle her book *Geographic Influences in American History*, but shortly before its publication, it was discovered that her near contemporary Albert Perry Brigham (1855–1932) was working on, and had copyrighted, a book of the same name. Retitled *American History and its Geographic Conditions*, Semple's book was published in 1903 – the same year as Brigham's. Despite their similar subject matter, Semple and Brigham proceeded 'as if the other one did not exist', and, as a consequence, drew fairly distinct conclusions.[175] Brigham had studied as a graduate student under Shaler and Davis at Harvard, and was familiar with Ratzel's work.[176] Where Semple's book took as its basis the influence of individual environmental factors upon the historical settlement of the United States (rivers, mountains, climate), Brigham took a rather more regional approach by considering the particular combination of geographical factors characteristic of specific 'physiographic provinces', and their subsequent influence upon national development.[177] As a consequence of the somewhat different emphasis placed upon geographical influences by Semple and Brigham, their work could be read as complementary rather than contradictory.[178]

In the critical response to the work of Semple and Brigham, it is possible to detect the influence of readers' distinct disciplinary and interpretative circumstances. The complex contours of this reception show a distinction between generally positive reviews by geographers and more critical interpretations by historians (the exception being Frederick Jackson Turner's review in *The Journal of Geography*). In demonstrating the validity and applicability of the anthropogeographical method to geographical work in the United States, Semple's text was welcomed both by American geographers – including Ralph Stockman Tarr (1864–1912), a former student of Davis – and international scholars, among them the Oxford geographer Andrew John Herbertson (1865–1915).[179] Semple's work, coming as part of the 'drama of professionalizing geography', was read by geographers as a contribution to debates then current regarding the infant discipline's epistemic and methodological foundation.[180] In this respect, Semple, despite holding no professional position within the academy, became 'part of the movement to establish a professional field of geography in America.'[181]

Something of the cultural and scholarly impact of Semple's volume is indicated by its relatively rapid adoption as a standard textbook on historical geography and anthropogeography. In a number of states, the book was adopted by Teachers' Reading Circles where it was read by elementary and secondary school teachers of geography, and was placed on the formal reading lists for history and geography at several universities.[182] Semple's book was also adopted by every ship's library in the United States Navy, and 'included in the list of required reading for students entering the government military at West Point.'[183] As a consequence of the comparative success of her volume, Semple became 'a person of importance who was in great demand.'[184] *American History* was, for Semple, an important watershed

– it marked the conclusion of her first period of geographical authorship, and, as a warrant of credibility, provided an entrée to the professional geographical community. As one Kentucky newspaper reported
of Semple's scholarly success, 'on a quiet street of a Kentucky city an
unassuming little woman ... [has produced] an authority for the
centuries to come.'[185]

Semple's professional apprenticeship

In 1904 the two most significant influences in Semple's life died: her
mother and Friedrich Ratzel. Shortly before his death, Ratzel expressed a desire that Semple should realise her long-held ambition of
communicating his anthropogeographical principles, in their fullest
expression, to the English-speaking world. He reassured her that there
was no one 'who could handle this matter as well as you.'[186] Driven
by the loss of her intellectual mentor, Semple enthusiastically pursued the task of translating Ratzel's ideas – beginning a seven-year
project that became *Influences of Geographic Environment* (1911).
Despite the personal losses experienced by Semple in 1904, her professional standing advanced considerably. In September of that year,
she was invited to present, along with Martha Krug Genthe (1871–
1945), a tribute to Ratzel before the Eighth International Geographical Congress in Washington, D.C.[187] Semple also addressed the
Congress' Educational Section on the pedagogical potential of anthropogeography.[188] The Congress was an important opportunity to
communicate to an international audience the geographical work
then being conducted in the United States, and the Chairman of the
Congress' Scientific Committee, William Morris Davis, was keen to
make 'the best possible showing.'[189]

Concomitant with Davis' international aspirations was the 'larger
problem of mobilizing geographers in the United States.'[190] For Davis,
the tasks of disciplining and professionalizing geography were related
and imperative. Whilst physical geography had begun to coalesce
under recognised courses, degree programmes, and (in some instances) departments at north-eastern universities in the last decade
of the nineteenth century, the same was not true of human geography – the very issue to which Semple's work was seen to speak.

Although two independent societies existed for the promotion of geographical knowledge – the American Geographical Society and the National Geographic Society – neither was tailored specifically to the requirements of newly-emergent academic geographers.[191] Davis was keen, therefore, to establish 'a society of mature geographical experts' that might more properly represent their interests.[192]

Davis' plan for an Association of American Geographers was given impetus by the 1904 Congress.[193] Before the close of the year, a seventy-strong list of potential members was compiled, based upon an evaluation of their published work. Of the forty-eight short-listed candidates who went on to become charter members of the Association, only two were female: Semple and Genthe. Unlike their male counterparts, neither Semple nor Genthe were 'employed in research oriented Universities.'[194] This gender discrepancy was typical of the period and belied the fact that Genthe was the only founding member of the Association to hold a Ph.D. in geography (from the University of Heidelberg). Davis was convinced of Semple's and Genthe's 'scholarly qualifications', and, in this respect at least, their gender was not an impediment to membership.[195] The wider organization of academic geography in the United States was such, however, that gender inequalities were significant and obvious.[196] Formal degrees were not considered, however, to be of crucial significance. One of the Association's founding members, Nelson Horatio Darton (1865–1948) had, for example, no university education.[197]

The Association, from its founding, comprised the leading American geographers and afforded Semple a warrant of professional credibility that complemented the positive reception of her scholarship. The Association's subsequent annual meetings provided her also with an important platform from which to communicate her ideas. Semple devoted much of 1905 (her first year as a professional geographer) to work on *Influences*, but took time away for a further visit to Europe. Whilst abroad, Semple was invited by John Scott Keltie (1840–1927), Secretary of the Royal Geographical Society, to lecture before the Society on either mountain dwellers or convict islands – subjects 'semi-popular and sufficiently narrow to be adequately treated in one evening's lecture.'[198] Semple was concerned to ensure that she might be permitted to extemporise during her talk rather than be compelled to read from a prepared paper – this despite what she described as 'an old-time feminine objection to hearing myself speak in public.'[199] Although Keltie was keen for Semple to lecture before the Society (and, indeed, also invited her to contribute a paper to *The Geographical Journal*), her plans changed and Semple returned to the United States without addressing the Society.

Upon returning to the United States, Semple was approached by Rollin D Salisbury (1858–1922), a fellow member of the Association of American Geographers, and founder of the newly-created department of graduate studies in geography at the University of Chicago, and offered a visiting lectureship.[200] Established in 1903 by Salisbury, then dean of the graduate school, the department was the first in the United States to offer graduate studies, and exerted an unequalled influence upon the development of the discipline during the first half of the twentieth century. Salisbury – regarded as a 'skilful organizer, an inspiring leader, [and] a teacher beyond praise' – assembled a faculty drawn from among the leading geographers of the period, including John Paul Goode and Harlan Harland Barrows (1877–1960).[201]

Salisbury had studied geology at Beloit College under Thomas Chrowder Chamberlin (1843–1928).[202] Salisbury, like his near contemporary Davis, pursued work in physiography (physical geography).[203] He conceived of geography as a geminated discipline, combining the related fields of 'geographic geology' and 'life-significance studies.'[204] The latter he understood to take as its focus the 'relevance of physical conditions to human affairs.'[205] This reflected a longstanding interest in environmentalist themes – cultivated during a rural boyhood in which Salisbury 'noticed interesting things on the farm', such as the influence of agricultural practice upon local topography.[206] In this respect, he was, like George Perkins Marsh, 'much more concerned with man's influence on nature than with nature's influence on man.'[207] Salisbury was, as a consequence, 'hopeful, but sceptical, that workers in anthropogeography might develop that part of geography on a scientific basis.'[208]

Something of Salisbury's interest in the promotion of anthropogeography was evident in his appointment to the department of Barrows who, in the summer of 1905, offered a course on the 'Influences of Geography on American History' – the first of a series of special Summer Quarter seminars intended for teachers of geography.[209]

Barrows' brand of environmentalism was, in common with Salisbury's vision, concerned with influence rather than with 'extreme determinism.'[210] In this context it might seem peculiar that Semple – later understood as an advocate of environmental determinism – should have been invited by Salisbury to join the faculty on a part-time basis. Her perspective on anthropogeography was not, however, logically incompatible with that of the department, even though Chicago would later be associated with a passionate 'revision of the environmental doctrine.'[211] Barrows' concern (and Salisbury's too) was to make clear the mutual relationship between society and environment.[212] Barrows drew, in this respect, on the work of Turner (under whom he later studied at the University of Wisconsin), and, most particularly, on Semple's *American History*.[213] Whilst Barrows' perspective on anthropogeography would later change it was, at the time of Semple's hiring, broadly compatible with her own.[214]

Semple's appointment to Chicago reflected not only the topicality of her geographical interests, but also the perceived authority of her scholarship. She was an important part of Salisbury's plan to 'break new ground in ... teaching and research.'[215] Her election to the Association of American Geographers undoubtedly facilitated this transition from private scholar to academic geographer. Her appointment to the department was on a part-time basis only (she lectured typically during the Spring Quarter). Quite why she was not offered, or did not accept, a fulltime position is uncertain. For one biographer, this was a conscious choice on Semple's part, and reflected 'her priorities in keeping research and writing as her primary activities.'[216] For Robert Swanton Platt (1891–1964), who studied under Semple at Chicago, Salisbury's decision to appoint Semple on a part-time basis was 'not so much because [he was a] nonbeliever [in] environmentalism as [it was] because [Semple was a] woman.'[217]

The course Semple developed – 'Some Principles of Anthropogeography' – was intended as a general introduction to her geographical perspective, drawing upon her existing body of work.[218] Semple's lectures were considered the 'most stimulating & inspiring' of those offered by the department, and her ideas were, as a consequence, received with considerable approbation.[219] The opportunity for Semple to present her work to an audience of enthusiastic students, the first in the United States to receive an explicitly geographical education at graduate level, proved valuable in shaping not only the subsequent content of her work, but also aspects of the discipline's later research focus. Semple's students included 'many who went on to play important roles in the development of professional geography.'[220] The influence of Semple and of Chicago were such that four fifths of geography Ph.D. graduates during or before 1946 could trace an academic lineage back to one of five geographers, four of whom taught at Chicago: Salisbury, Barrows, Semple, Wallace Walter Atwood (1872–1949), and Vernor Clifford Finch (1883–1959).[221]

THE GENESIS OF *INFLUENCES*

Semple's relatively light teaching schedule ensured that she was able to devote extended periods to her work on *Influences*. When not teaching in Chicago, she divided her time between Louisville and the Catskill Mountains in New York State, where she lived in a tent and worked on her book without interruption. As one newspaper reported, 'she would work six hours a day, with only the chipmunks and the birds as her companions.'[222] Rather than present a literal translation of Ratzel's *Anthropogeographie*, Semple sought to re-examine the fundamental principles of his work, to clarify them, to subject them to proof and, where necessary, to reject them. She intended to relocate Ratzel's book linguistically, and to reframe its contents, revise its arguments, and supplement its sources. She sought to 'make the research and induction as broad as possible, to draw conclusions that should be elastic and not rigid or dogmatic ... to be Hellenic in form but Darwinian in method.'[223] Semple hoped to distinguish her text from Ratzel's in several ways. The first was to eliminate the organic theory of society and state, which had formed an important interpretative component of Ratzel's work. Additionally, Semple was disinclined to use race as an explanatory category, believing that if people of different ethnic stock, but similar environments, manifested similar or related social, economic, or historical development, it was reasonable to infer that such similarities were due to environment and not to race. Perhaps most significantly, however, Semple's explicit aim was to deny any straightforward relationship between the natural environment and human social and physiological organization.[224] Her prefatory remarks made this clear: 'the writer speaks of geographic factors and influences, shuns the word geographic determinant, and speaks with extreme caution of geographic control.'[225] In this respect, although it was Semple's intention to offer a 'faithful English rendition of her master's doctrine', her project was one of reinterpretation, rather than straightforwardly of representation.[226]

By mid-1907 Semple had made substantial progress on *Influences* and was keen to communicate her findings. In addition to two papers in the *Bulletin of the American Geographical Society* dealing with geographical boundaries, Semple was invited by Frederick Jackson Turner to contribute a paper to the meeting of the American Historical Association in Madison, Wisconsin.[227] Turner's session – to which he offered a paper on 'The relation of geography and history' – was intended to bring together work on geography, history, and environmental influence. Semple's paper, 'Geographical location as a factor in history', was not, as has been claimed, 'the first occasion on which Semple delivered a formal paper before an assembled body of scholars', but was nevertheless an important opportunity to communicate her ideas to an interdisciplinary audience.[228]

Semple's work was not received with unequivocal enthusiasm. One member of the audience – George Lincoln Burr (1857–1938), then professor of medieval history at Cornell University – took exception to aspects of Semple's thesis, and engaged her in extended debate. His principal contention was that Semple placed too much emphasis on geographical control. For Burr, 'geography, though a factor in history, is only a factor, and that no more in history than in mathematics can the outcome be inferred from a single factor alone.'[229] Attempting to strike a conciliatory note, Harlan Barrows, who was also in the audience, 'defended a position intermediate between that of Miss Semple and Professor Burr.'[230] Semple for her part was inclined to attribute Burr's dubiety to the fact that, as she perceived it, 'historians as a rule do not know geography.'[231] She was supported in this opinion, to some extent, by Ralph Tarr (who had earlier praised her *American History*) and George Burton Adams (1851–1925), President of the American Historical Association, who believed 'the disagreement was caused partly by lack of definition of terms.'[232] The argument resurfaced some days later at the Association of American Geographers meeting in Chicago, where Semple and Burr discoursed at length during the formal dinner.[233]

Throughout the research and writing of *Influences*, Semple maintained a correspondence with John Scott Keltie. On 21 April 1907, she dispatched a paper dealing with coastal peoples (which later formed the eighth chapter of *Influences*), along with a letter outlining her approach to, and hopes for, the book. This letter indicates Semple's eagerness to ensure that her work was seen as 'something more than a mere restatement of Ratzel's principles.'[234] To that end, Semple had 'made wide inductive research, just as if I were writing a wholly original work', enabling her, she believed, to see 'more clearly than he [Ratzel] did ... the immense importance of the *interplay* of geographic forces.'[235] Impressed by Semple's chapter, Keltie again invited her to lecture before the Society. She agreed enthusiastically, hoping that she would 'finish the manuscript and maps a year from this date' and then talk 'volubly on the subject of Anthropo-geography' at the Society.[236]

Between 1908 and 1910, Semple maintained an almost unwavering pattern of research, writing, and presentation as *Influences* took shape. Her 1907 talks at the American Historical Association and Association of American Geographers were published the following year in the *Bulletin of the American Geographical Society*, and on 27 November 1908 she presented 'The operation of geographic factors in history' at the annual meeting of the Ohio Valley Historical Association in Marietta, Ohio.[237] Semple's paper formed the basis of the first chapter of *Influences* – indeed the opening sentence of both was identical: 'Man is a product of the earth's surface.'[238] In 1910, perhaps in anticipation of the imminent publication of *Influences*, Semple's then famous paper on the Kentucky Mountains was republished in the *Bulletin of the American Geographical Society*.[239] Semple had been 'constantly getting requests' for offprints, and the copy on deposit at the library of the University of Chicago had been used so heavily that the 'article has the printer's ink almost now off.'[240] By the close of the first decade of the twentieth century, then, Semple's work on *Influences* neared completion and its potential audience had been alerted to its publication through papers and conference contributions.[241]

In March 1910, Semple wrote to Keltie from the University of Chicago, announcing that her manuscript was nearing conclusion. It is evident from her letter that she intended *Influences* to meet the requirement of student geographers: 'I have had the advantage of lecturing out the material three times here at the University of Chicago; and this has enabled me to adapt it to students' needs.'[242] Semple's principal reason for writing to Keltie was to gauge his view on the suitability of issuing a British edition of *Influences*. She wrote: 'Do you think it would perhaps be advisable to arrange for an English edition of it [*Influences*], in view of the growing demand for geography in your universities? I should greatly appreciate a word of advice from you … for no one understands the English field so well as you.'[243] In reply, Keltie enthused:

> I shall be very interested indeed to see your book on the Influences of Geographical [sic] Environment, when it is published. We want a book which discusses the whole problem thoroughly, widely and fully. We talk a great deal about the influence of geographical environment, but I do not think that anyone has actually and fairly faced the position, stating what the terms of the problem are on both sides, first from the side of the environment – what exactly do we include in that term; and then from the side of the human subject, and what precisely as far as we can make out, are the interactions between them. I should very much like indeed if an English publisher would take the book up.[244]

Although Keltie advised Semple to discuss this matter with her publisher, Henry Holt and Company, he suggested a number of suitable London firms, including Macmillan, Heinemann, and John Murray.

When Semple next wrote to Keltie, on the eve of the publication of *Influences*, it was in a mood both buoyant and reflective. Contemplating her recently-completed work, she explained:

> I hoped to make the research and induction as broad as possible, to draw conclusions that should be elastic and not rigid or dogmatic, and finally to give the whole book a certain literary quality …. That was my ideal: of course I did not get within shouting distance of it in the accomplished book, as you will clearly see; but perhaps you will occasionally catch a gleam from the star to which I tried to hitch my lumbering little cart.[245]

This apparent lassitude was countered by the enthusiasm she expressed for a planned round-the-world journey:

> now I'm to have my play time; early in June I start on a year's trip around the world viâ San Francisco and Japan .... Some time in the summer or autumn of 1912, I shall loom up on the horizon of Burlington Gardens; there I shall drop into the house of the Society and say, – how do you do, Dr Keltie, do you remember me?[246]

Keltie's reply, congratulating Semple on the publication of *Influences*, reached her shortly before she departed on her global sojourn. In it, Keltie's expectancy is evident:

> I am delighted to hear from you once again, especially with such good news about your new book. We have not received it yet, but I dare say we shall soon, and you may be sure that I shall read it with real delight, and hope we shall be able to have a stunning review of it in the Journal by some competent hand.[247]

CONCLUSION: THE DEVELOPMENT
AND PROMOTION OF ANTHROPOGEOGRAPHY

Semple's *American History* included on its title page the Ritterian epigram 'So much is certain: history lies not near but in nature.'[248] In some respects, this quotation characterized Semple's approach to geography – one which saw the natural environment as 'the central determinant' of history.[249] Anthropogeography was, for Semple, a method which combined her dual intellectual interests: history and geography. It was also an approach which, for a period of almost four decades, paralleled and reflected the scholarly concerns of disciplinary geography – first in Germany, then later in the United States and Britain. Although part of a broader contemporary debate concerning evolution, nature, society, race, and cultural development, anthropogeography represented a nineteenth-century revision of Classical environmentalist themes.

Ratzel's formulation of anthropogeography – having drawn upon the earlier work of, among others, Ritter, Peschel, and von Richthofen – can be seen to be part of a tradition within German geography concerned with describing the relationship between people and land. Yet it might equally be seen to depend upon the emergent themes in nineteenth-century biology and ethnography, particularly that of Darwinian evolution and its subsequent social revisions. The genesis of Ratzel's anthropogeography can be seen most properly to represent the unique combination of these scholarly themes. His intellectual interests were informed not only by his academic mentors, including Haeckel and Wagner, but also by his periods of journalistic field observation as correspondent for the *Kölnische Zeitung*. Ratzel's pesonal biography mattered, then, to the development of his academic concerns.

Personal experiences mattered also to the formation of Semple's geographical interests. Firsthand experience of the Civil War in Kentucky; of class segregation in Louisville; and of extreme poverty in the Appalachians were motivating factors in the development of her environmentalist concerns. Semple's privileged upbringing also facilitated the educational and research opportunities which engendered her dual interest in history and geography. When considered in light of her subsequent research focus, her period under the tutelage of Ratzel was the most crucial pedagogic experience. In Ratzel's work, particularly as communicated in the first volume of his *Anthropogeographie*, Semple found an expression of geography which corresponded with her own emerging perspective. Although an enthusiastic and faithful student of Ratzel, Semple was not an uncritical disciple. In communicating Ratzel's geography to the Anglo-American academic community, she saw an opportunity to correct its perceived failings by putting it on a more rigorously scientific foundation, based principally upon field observation.

Semple's promotion of a scientific approach to geographical research coincided with the institutionalization of the discipline in the United States. Her interpretation of Ratzel's ideas was seen to correspond not only with a desire among geography's proponents to place the discipline on a scientific footing, but also with the neo-Lamarckian environmentalist project outlined by, among others, Turner, Marsh, and Shaler. It was the topicality and applicability of Semple's early work on anthropogeography that ensured its generally positive reception. Yet it was not until Semple entered the academy in her early 40s that she felt sufficiently able to begin the major part of her geographical work: the writing of *Influences*. Semple's appointment to the University of Chicago afforded her the opportunity to communicate her ideas to an enthusiastic graduate cohort and to work through the principal components of her book in the lecture theatre. This space, a forum for debate and discussion, was one in which Semple's oratorical skills were honed and her philosophy refined. The publication of *Influences* in 1911 was, in certain respects, the apotheosis of Semple's anthropogeographical project. Rather than marking the terminus to this particular element of her research, however, the book was a prompt to a new and important phase of geography's disciplinary development. In the chapter which follows, I trace the initial reaction to Semple's book through an examination of reviews in the popular press and academic literature, and describe how these early readings framed the response to *Influences*.

3 Popular and scholarly reviews of *Influences*

Common to work in the history of science and literary criticism is the idea that reception includes the afterlife of an initial textual encounter. Rather than being simply a temporally-fixed event – the moment when the reader scans a line of text and begins to consume or construct its meaning – reception is also what happens next. The reception of *Influences of Geographic Environment* was, then, a question not only of its initial reading, but also of the role of anthropogeography in informing then-current discussions in geography; of its incorporation into teaching curricula; and of the subsequent rejection of the geographical perspective it sought to convey. In thinking about the trajectory of Semple's anthropogeography, or the career of *Influences*, it is necessary to consider what her book meant to its various audiences in 1911 (and why), and also what it meant to readers at various times, and in different places, in the years following its publication. The study of reception is, in this way, 'concerned with investigating the routes by which a text has moved and the cultural focus which shaped or filtered the ways in which the text was regarded.'[1] Reception is, then, a process that is never completed – a series of moments described by individual reactions and encounters, unconstrained by an arbitrary limit of time.

Efforts to reconstruct the reception of scientific knowledge have, as was described in Chapter 1, tended to employ in their analyses a hierarchical conception of scale, privileging the national.[2] An unintended limitation of this approach is a propensity to de-emphasize interpretative differences *within* nations, whilst accentuating dissimilarities *between* them. The assumption of homogenous analytical practices which underpins the examination of national responses to science has been challenged by work which has attended to local and individual responses to scientific and theological texts.[3] As might be expected, however, examining reading and reception practices at the level, variously, of region, city, street, or individual reveals an apparently unbounded heterogeneity of interpretation: the more local the scale of analysis, the more diverse and particular the hermeneutic practices appear to be. In short, the closer we look, the more we see. If the intention of work in the reception of texts and of knowledge is to make claims about the nature of circulation and consumption which are in some way general and look beyond individual experience, then it is necessary to consider in what ways commonalities and shared interpretations might be identified. As has been suggested, 'Precisely what the correct scale of analysis is at which to conduct any particular enquiry into the historical geography of science – site, region, nation, globe – has to be faced.'[4]

Given the fact that at different scales, different patterns of reception can be identified, the utility of spatial scale as a framework upon which to reconstruct the reception of knowledge is uncertain. Although the intention of this book is not to challenge the ontological or epistemic validity of scale as a basis to describing the reception of Semple's anthropogeography, it is apparent that to better understand the commonalities and disunities in the response to *Influences*, it is helpful to think beyond scale as nominally-fixed, spatially-defined categories, and to consider the function of social networks and hermeneutic communities. To speak of the reception of Semple's work it is necessary to consider what reception means, and at which spatial, temporal, and social scales such meanings can usefully be explored.

The popular and professional reviews of *Influences* are one way in which to interrogate the reaction to Semple's text and to investigate the value of social networks and spatial scales in accounting for the book's reception. Ranging from the highly complimentary to the mildly derogatory, the printed critiques of *Influences* reflect the spectrum of opinion associated with its reception, but are not straightforwardly a proxy of it. Indeed, as one historian of science has noted, 'reviews are by no means the only standard by which reception and relative success of books can be measured.'[5] Moreover, given the general anonymity of the newspaper reviews upon which this chapter draws in reconstructing the popular interest in *Influences*, it is not always possible to determine the identity of individual readers, or, when their identity is apparent, to contextualize their reading experience and to position them within an interpretative community (other than that of the audience which the newspaper addressed or aspired to address).

The professional reviews of Semple's book – those which appeared in learned journals and academic periodicals – present other, related analytical difficulties. Although the identity of the reviewers in these cases is almost always apparent, quite how their interpretation of *Influences* was conditioned by their disciplinary concerns, and those of the periodical for which they were writing, is not always clear. The extent to which these reviews can be seen to represent discrete disciplinary reactions to Semple's book is limited, moreover, by the same vagaries of individual authorship, editorial remit, and reviewing culture evident in relation to the popular assessments of *Influences*.[6] Generally speaking, however, it might be assumed that the authors of these professional reviews were addressing audiences who shared certain academic concerns, who were familiar with particular canonical texts and debates, and who wished to know the value of Semple's book in relation to their own disciplinary context.

Rather than framing my analysis of the popular and professional reviews of *Influences* in relation to their thematic content or geographical origin, this chapter follows a more-or-less chronological arrangement. My intention in so doing is not to present an uncomplicated narrative of the book's reception, but simply to acknowledge the thematic complexity of newspaper and periodical reviews and the difficulty inherent in their categorization. Given that the majority of British newspaper reviews were published after those in the United States, and that periodical reviews were issued later still, a coincidental grouping of location and medium is evident. Whilst these accidental categories provide more opportunity to make general claims about the role of location and publication type in the reviewing of Semple's book, they are not intended to provide definitive conceptions of the British, or American, or newspaper, or academic response to *Influences*.

Reception is messy and various and personal – reviewing no less so. Whilst this complexity precludes a definitive narrative of the critical response to Semple's work, it invites useful speculation about the nature of interpretation, the geographical variability of reading, and the processes which facilitate the communication and reception of knowledge. The discussion and assessment of Semple's book by what might be thought of as a republic of reviewers, served to define the initial trajectory of *Influence*'s reception – acting as an interpretative buffer between Semple and her intended audience. For that reason, as this chapter seeks to do, it is important to consider the nature of those reviews; the factors which shaped their authors' reading of *Influences*; and what it is that these critical responses can tell us about the reaction of different interpretative communities to Semple's book. In tracing the critical reading of *Influences*, I reflect on the role of medium and audience in determining the culture of reviewing, examine how different intellectual and disciplinary concerns influenced the book's perceived value and credibility, and consider quite how we might deal with the geography of reviewing practice. Enumerating these reviews is not an exercise in comprehensiveness; rather it is an attempt to illustrate how, from a heterogeneous collection of individual reading experience, it is possible to identify interpretative commonalities.

THE NATURE OF *INFLUENCES*

In a brief preface to *Influences*, Semple detailed her perspective on the book's function and purpose and outlined the intellectual genesis of her anthropogeography. Her intention was to show that, although planned originally as a 'restatement of the principles embodied in Friedrich Ratzel's Anthropo-Geography', her book had developed to become something rather more sophisticated and intellectually relevant, reflecting the geographical interests of its author and mirroring contemporaneous disciplinary concerns.[7] As a consequence of her earlier discussions with John Scott Keltie, Semple was keen to make clear to her readers that, in bringing *Anthropogeography* to the United

States, her project was not one merely of 'literal translation', but rather was an exercise in interpretation and cultural relocation.[8]

It was important for Semple that *Influences* should be 'adapted to the Anglo-Celtic and especially to the Anglo-American mind.'[9] The purpose of this cultural reframing was to place Ratzel's work more obviously on a scientific foundation and to 'throw it into the concrete form of expression demanded by the Anglo-Saxon mind.'[10] Semple's concern was, as it had been in her earlier work, to reform Ratzel's conclusions, which she regarded as 'not always exhaustive or final', and to present them in a manner more clearly supported by real-world examples.[11] In so doing, Semple drew upon 'about a thousand different works' – bringing together data from travel and exploration texts, and from 'works of comprehensive or even encyclopedic scope in the fields of history, geography, and anthropology.'[12] Her desire to situate her work within this literature represented her wish to position anthropogeography in relation to a wider intellectual genealogy and also a pragmatic attempt to avoid the 'just criticism of inadequate citation of authorities' to which Ratzel had been subject.[13] *Influences* combined and juxtaposed contemporary sources with Classical authorities. In uniting such disparate work, Semple's sought to 'compare typical peoples of all races and all stages of cultural development, living under similar geographic conditions.'[14] If, by so doing, she was able to show that 'peoples of different ethnic stocks but similar environments manifested similar or related social, economic or historical development', she might convincingly make the inference that such similarities were a function of environment rather than of race.[15] Semple was aware, however, of the potential speciousness of this argument, and felt compelled to state that she had 'purposely avoided definitions, formulas, and the enunciation of hard-and-fast rules' in describing the causal links between environment and society.[16] The purpose of *Influences* was, she stated, not to 'delimit the field' or to advance 'precipitate or rigid conclusions', but was to serve as an indicative manifesto for what anthropogeographical research was and might become.[17]

The first chapter of *Influences* – 'The operation of geographic factors in history' – opens with a Scripturally-resonant proclamation: 'Man is a product of the earth's surface.'[18] This statement underpins Semple's proposition that "man" (to employ her term) cannot be studied scientifically, or understood correctly, without consideration being given to 'the ground which he tills, or the lands over which he travels, or the seas over which he trades.'[19] The body of the first chapter is devoted, therefore, to a wide-ranging summary of human/environment interactions in historical context. In a series of case examples, drawn from her wider reading, Semple outlines her perspective on various components of geographical influence (topographical, climatological, geological, hydrological, among others), describing the different ways in which these factors have affected human society, psychology, and physiology. Despite her noted desire to speak of geographical influence rather than of geographical determinant, the section dealing with climate attributes to it controlling influence on aspects of human life: 'Climatic influences are persistent, often obdurate in their control.'[20]

In its second chapter, *Influences* details at greater length the classes of geographical influence previously identified. Here, again, Semple's tone is rather more deterministic than might be expected given her protestations against this line of argument. The text is peppered with language incompatible with her desire to avoid 'the word geographic determinant.'[21] She speaks in terms of the '*pressure* of the environment', and about the ways in which the 'environment modifies the physique of a people ... by *imposing* upon them certain dominant activities.'[22] Humanity is described as 'a *passive* subject', exposed to environmental factors that '*determine* the direction' of its development, and '*determine* the size of the social group.'[23] The reason for Semple's use of such seemingly inconsistent language is not immediately apparent, but 'such adverbs as "inevitably," "always," and "everywhere" are favoured' in a way that is at odds with 'avowals to the contrary in the preface.'[24] A similar tone pervades the following fifteen chapters, and is apparent in her discussion of all aspects of human/environment interaction.

READING THE POPULAR RECEPTION OF *INFLUENCES*

On 11 March 1911 in its 'Book news and book views' column, the Syracuse, New York newspaper *The Post-Standard* reported the imminent publication of Semple's book. *Influences* had been selected for special mention from among Henry Holt's March output, but the anonymous *Post-Standard* copywriter seems to have been unfamiliar with its author; Semple is introduced incorrectly as 'Ellen Church Temple.'[25] Beyond *The Post-Standard*'s erroneous two-sentence advanced notice, *Influences* did not attract further press attention until June (for reasons that remain unrecorded, the book's publication was delayed until 29 May).[26] After this apparent false start, a comparative flurry of publicity accompanied the book's summer launch – *The Nation* (New York City), for example, carried seven advertisements between June and September. Despite the effort the book's production had cost, Semple chose not to await critics' reaction and embarked almost immediately on an eighteen-month journey around the world. Aware of the significance of newspaper reviews in framing the reception of her work, however, Semple had these collated by a press clipping bureau and dispatched to her at intervals during her sojourn. She was keen to learn whether these reviews foresaw 'the career for the book which I had hoped for.'[27]

A brief and matter-of-fact summary appeared on 17 June in *The Publishers Weekly*, a New York City trade news magazine serving the publishing industry, booksellers, and librarians.[28] Semple was described simply as 'The author of "American history and its geographic conditions"', and her book as 'a modified, simplified, and clearer ... restatement of Ratzel's "Anthropo-Geographie".'[29] Rather more contextual exposition was found in the following day's *Daily Picayune*, a New Orleans newspaper. The *Picayune* had formerly been a powerful organ of pro-slavery politics, and was still in 1911 largely 'white, conservative and racist.'[30] The political stance of the paper was in flux, however, and was becoming more closely aligned with the Democratic position of its rival, the *Times-Democrat*, with whom it merged in 1913. Given the important role of race and geography in the historical development and contemporary politics of New Orleans, it seems likely that Semple's exploration of the topic would have been of some interest to the *Picayune*'s 28,600 readers.[31] As with the notice in *Publishers Weekly*, it was assumed by the *Picayune* that Semple was 'already well known' to its audience as the author of *American History*. *Influences* was described as an extension of the themes outlined in that book, namely 'How geography goes hand in hand with history and sociology.'[32] For readers unfamiliar with the scope of anthropogeography, its purpose was defined as being to show how 'social and historical development has been affected by such factors as climate, soil, rivers, seas and mountains.'[33]

In contrast to the rather perfunctory announcements of *The Publishers Weekly* and the *Daily Picayune*, a more thoughtful and considered response to Semple's book featured in *The Sun* (New York City) on 24 June.[34] The anonymous reviewer seems to have taken as the basis of his or her response Semple's hope that her book would fulfil a pedagogical role. Under the headline 'Geographical light on history', *The Sun* lamented the current state of school geography, noting that whilst children 'are taught about climate and physical configuration, about the place of the earth in the universe, about nature and strange peoples ... they do not know that Springfield is in Massachusetts or the Ozarks in Missouri.'[35] Although disapproving of the general trend towards specialization in geography, the reviewer was keen to make clear the value of Semple's work: 'none [of geography's specialisms] is so fascinating as the "anthropogeography" of Katzel [sic] and Peschel which Ellen Semple Churchill [sic] introduces.'[36] The reviewer saw much in Semple's book that engaged with 'the study of plain geography', something which was understood to have been among 'the greatest sufferers in the evolution of the modern school system.'[37] In its attention to geographical context and to environmental circumstance, Semple's work was seen by this reviewer to incorporate fundamental components of a correct geographical education.

Beyond the empirical content of *Influences*, the review found Semple's causal scheme linking human history to its geographical situation largely valid. So convincing were the examples Semple advanced, the reviewer expressed the ironic concern that they risked eliminating 'pride of individual achievement or national characteristics' by showing geographical location, rather than a population's inherent merit or ability, to be the controlling factor in its social development.[38] Despite the reviewer's support for Semple's specific claims, he or she expressed some dubiety about their wider applicability. As the reviewer framed it, 'The danger with the science [anthropogeography] is that, while the theories may be true and may be applicable in general cases, in specific instances other elements also come into consideration.'[39] The reviewer's concern was that although Semple's principles were confirmed by the specific examples she provided, their general applicability was unproven – particularly when they were formulated and proposed as scientific rules. The reviewer recommended treating Semple's work not 'as an exact science, but as a tentative explanation of many things that have happened on earth.'[40]

The Sun's review concluded by stating that Semple 'has rendered education a service' by expounding upon a component of geographical research 'which the modern pedagogues are inclined to neglect.'[41] Her repeated reference to German and French authorities was singled out for particular praise, as was her 'interesting and readable manner.'[42] By situating her work in relation to its perceived intellectual genealogy and by applying to her writing the literary style which she and Ratzel believed necessary for the communication of anthropo-geography, Semple secured the approbation of The Sun's reviewer. A doubt remained, however, about the scientific validity of her work, particularly in terms of its ability to furnish nomothetic propositions. As a consequence, the reviewer concluded that Influences 'is an admirable piece of work, provided it is not used as a text book.'[43]

The Sun's generally laudatory assessment of Influences was echoed by the Boston Evening Transcript on 5 July.[44] The Transcript had something of a tradition of printing items relating to current debates in geography. In 1849, for example, it had given considerable attention to the public lectures in Boston of the Swiss geographer Arnold Guyot, and, towards the end of the century, it regularly included letters and notes from William Morris Davis.[45] As with previous reviews of Influences, and perhaps as a consequence of its earlier geographical output, the Transcript assumed in its readers knowledge of Ratzel and his work. The precise details of Semple's thesis were not, consequently, made immediately obvious: her book was described only as being 'on the basis of Ratzel's monumental system of anthropo-geography.'[46] The reader of the Transcript was to understand that it was as an addition to, and correction of, Ratzel's work that Semple's book was most valuable: 'The ideas of Ratzel have been tested and verified and the author in her work has had constantly in mind the English-reading peoples for whom her work has been prepared.'[47] In common with The Sun, the Transcript's reviewer thought Semple's 'extended reference to books and personal authorities' was important in extending the value and credibility of her conclusions.[48] Unlike *The Sun*, however, the *Transcript* thought *Influences* had more than simply a pedagogical value; it was seen to have a national importance and to be a 'distinct credit to American scholarship.'[49] For this reason, and as a consequence of Semple's accessible prose, the reviewer felt that *Influences* was likely to appeal both to 'the special student and to the general reader.'[50]

Quite who the general reader of the *Boston Evening Transcript* was in 1911 is an interesting question. The *Transcript* was Boston's foremost newspaper and attended particularly to the city's art and literature.[51] The paper's literary editor, William Stanley Braithwaite (1878– 1962), encouraged the work of emerging poets, one of whom, T. S. Eliot, immortalised the paper in his 1915 poem 'The *Boston Evening Transcript*'.[52] In Eliot's poem, the *Transcript*'s readership was set apart from the city's lascivious and sanguine street life: 'When evening quickens faintly in the street, / Wakening the appetites of life in some / And to others bringing the *Boston Evening Transcript*.'[53] The *Transcript*'s readers were the 'local intelligentsia and upper social class' of Boston and Cambridge, and one might assume that their literary sensibility and familiarity with certain geographical debates (a consequence of Davis' contributions), would have facilitated a critical engagement with Semple's text.[54]

It is unwise to assume, however, that the general reader as imagined in retrospect is the same general reader to whom the *Transcript*, or the *Picayune*, addressed their reviews of Semple's book. Whilst is it possible to infer something of the interpretative stance of a newspaper's audience from its social characteristics – 'intelligence, socio-economic status, occupation, educational level, and so forth' – its inherent heterogeneity means that neither the *Picayune* nor the *Transcript* can stand as an unproblematic proxy for their various audiences.[55] The 'active plurality' of a newspaper's readership is such, moreover, that it would be unjustified to make broader inferences about the intellectual and hermeneutic characteristics of their readers' metropolitan setting based solely upon single reviews.[56] The *Transcript* alone cannot be seen to represent Boston's reading public.

Modes of reviewing and the problem of anonymity

The Outlook, a weekly New York City periodical, whose contributors included former United States President Theodore Roosevelt (1858–1919), published a chiefly complimentary review of *Influences* in its 15 July issue. Although the review was unsigned, the tantalizing possibility exists that Roosevelt was its author. Roosevelt had a long-standing interest in Social Darwinism, and his account of the United States' westward expansion during the eighteenth century – *The Winning of the West* (1889–1896) – formed an important basis not only to Frederick Turner's frontier thesis, but was also cited in both Semple's *American History* and *Influences*.[57] For *The Outlook*, Semple's text represented a 'valuable and scholarly' contribution to geography – one likely to prove 'of genuine interest to a considerable class of intelligent general readers.'[58]

In common with *The Sun*, however, *The Outlook*'s reviewer believed that the general applicability of Semple's anthropogeographical principles had been overstated: 'sometimes too much is claimed for the effect of geographic conditions upon man's development.'[59] Despite this qualification, the reviewer saw Semple's book as a valuable corrective to Houston Stewart Chamberlain's (1855–1927) *Die Grundlagen des neunzehnten Jahrhunderts* (1899), published in English in 1911 as *The Foundations of the Nineteenth Century*. Chamberlain's text argued for the controlling influence of race upon social development and advocated the desirability of preserving the Aryan race, whom he saw as the inheritors of Classical civilization.[60] Roosevelt later offered a withering review of Chamberlain's book in his 1913 volume *History as Literature*. For Roosevelt, Chamberlain's doctrine was 'based upon foolish hatred', and situated in 'a matrix of fairly bedlamite passion and non-sanity.'[61] For *The Outlook* and, one might suppose for Roosevelt also, Semple's expressed desire to eliminate 'the race factor' enhanced the value and relevance of her book.[62]

The Outlook's laudatory sentiments were echoed by the *Providence Daily Journal*, whose congratulatory assessment of Semple's book featured under the headline 'A German dose sweetened'.[63] In common with previous newspaper reviews, the *Journal* considered Semple's principal achievement to have been shaping Ratzel's *Anthropogeography* – 'a German work said to be difficult reading even for Germans'

– into a form accessible by the 'English and American students' to whom its contents had previously been largely unavailable.[64] Despite expressing some concern as to the limitations of Semple's conclusions, the *Journal* was satisfied by her admission that 'some of the principles may have to be modified or their emphasis altered after wider research.'[65] This minor caveat did not 'detract from the interest of the elaborate work, which shows the science as it is to-day, and which contains much that appeals to the intelligence and judgement of the thoughtful reader.'[66]

The *Journal*'s review, taken together with those published in other newspapers during the summer of 1911, demonstrates the general approbation with which Semple's ideas were greeted in the United States. A common source of praise was Semple's academic rigour and literary flourish. For the *American Library Association Booklist* (Chicago), for example, *Influences* was distinguished by 'Sound scholarship and a readable style.'[67] More particularly, however, Semple's achievement was seen to lie in 'liberating anthropo-geography from the drag-weight of the "social organism" theory of society', and in placing it more properly in the context of current social theory.[68] For the *Boston Herald*, so impressive was Semple's reformulation of Ratzel's work that *Influences* could, conceivably, 'be advantageously re-translated for the use of Germans themselves.'[69] For the *Herald*, then, the fundamental correctness of anthropogeography was not in doubt, and the value of Semple's book depended upon her proper framing and contextualisation of its principles. By presenting her conclusions with 'modesty and reserve', and by making clear their 'merely tentative character', it was the *Herald*'s opinion that Semple 'increases rather than diminishes the value of her book.'[70]

The relatively uncritical acceptance by the *Boston Herald* of Semple's anthropogeography contrasted with the thoughtful and measured assessment offered by the *Springfield Daily Republican*. Established in 1824 as a weekly Whig newspaper, the *Republican*'s political stance changed during the mid-nineteenth century as it became an important opponent of slavery.[71] In its Liberal Republican guise, the *Republican* was the political antithesis of the New Orleans *Daily Picayune*. Whilst both newspapers welcomed Semple's book, it seems probable that they did so for different reasons. Unlike previous assessments of Semple's work – which dealt only with the broad tenets of her environmentalist position – the *Republican* attended to her pronouncements on the geographical regions with which its readership would be most familiar: the continental United States, and, most especially, its eastern seaboard. By addressing the local application of anthropogeographical principles, rather than treating them in abstract, the *Republican*'s reviewer presented a geographically-specific assessment of their validity.

Although largely appreciative of the book's philosophy, the *Republican* expressed concern as to Semple's discussion of climate and its effect on different social groups. As the reviewer made clear, 'The eastern coast of the United States gives a specially [sic] good opportunity for the study of climate and its influence upon man' since, in a relatively narrow latitudinal range, the climatic variations are pronounced.[72] According to Semple's scheme, the 'contrasts in temperament, manner of life, point of view, etc.' resulting from this climatic gradient should be particularly marked.[73] Semple cites, by way of explanation, 'the famous contrast between New England Puritan and Virginian Cavalier', and concludes that the divergent population characteristics of the Northern and Southern states have 'become still more different owing to the fact that the large negro labouring class in the South, itself primarily a result of climate, has served to exclude foreign immigration.'[74] This conclusion represented, for the *Republican*, 'too strong a statement', in part because Semple had failed properly to acknowledge the 'French Huguenot and Scotch-Irish settlement' in Southern states.[75]

Despite the tentative nature of Semple's claims as to the significance of climatic influence, the *Republican*'s reviewer felt that she ought to distinguish more overtly 'between the direct and indirect effects of climate.'[76] Semple was, in fact, well aware of the importance of so doing. She thought it vital to 'distinguish between direct and indirect results of climate, temporary and permanent, physiological and psychological ones, because the confusion of the various effects breeds far-reached conclusions.'[77] She was conscious, moreover, that the 'direct modification of man by climate is partly an *a priori* assumption' and that 'incontestable evidences of such modifications are not very numerous.'[78] Despite this circumspection, Semple was explicit in her view that whilst the direct physical effect of climate was not always obvious, it 'undoubtedly modifies many physiological processes' and informs a population's temperament and intellectual energy.[79] The *Republican*'s reviewer broadly concurred with this position, and, in what was certain to appeal to the newspaper's local readership, summarised Semple's interpretation thus: 'The northerner is more domestic, and works harder; the southerner is less thrifty and feels less compulsion to work. Hence class lines are sharper in the South because in the North the labourer, under the whip of climate, is constantly recruited into the rank of capitalist.'[80] Semple's representation here of Jedidiah Morse's (1761–1826) moral topography (as articulated in his 1789 volume *The American Geography; or, a View of the Present Situation of the United States of America*) cast the *Republican*'s readership in a positive light – an accidental compliment which flattered the reviewer.[81]

Despite the largely positive response to *Influences* evident in its early newspaper reviews, efforts to assess the function of Semple's book – beyond its pedagogical role in informing the 'thoughtful reader' or 'student of anthropology' – were somewhat limited.[82] In almost all cases, reviewers assumed or implied that their readers 'had favourable [prior] knowledge' of Ratzel's *Anthropogeography* and Semple's *American History*.[83] Framed in this way as part of an already-established body of knowledge, and as an 'index to Ratzel's thought', *Influences* required a somewhat lower burden of proof than might otherwise have been expected.[84] Whilst the scholarly authority of Semple's book depended upon her attempts to place anthropogeography on a scientific basis, its popular authority was a function of her 'direct transmission of Ratzelian principles into American cultural and geographical understanding.'[85] The fundamental validity of Semple's thesis seems not to have been at question.

It is possible, conversely, that the apparent reluctance of newspapers' reviewers to assess the content of *Influences* in an analytical manner was a consequence of their very unfamiliarity with its principles. As one biographer of Semple has noted, anthropogeography was 'new to the United States, and few critics felt competent to deal with such a theoretical work.'[86] The authority and credit which they sought to confer upon Semple's book as a proxy of Ratzel's anthropogeography might equally represent an effort not to appear ignorant. The popularity of Semple's *American History* would suggest that a significant proportion of *Influences*' reviewers were likely to have read it, or have had access to it. Ratzel's *Anthropogeographie*, by contrast, was a 'closed book': its language and relative unavailability in the Anglophone world being limiting factors.[87] The familiarity with which Ratzel's work was treated in these reviews is, perhaps, erroneous, and illustrative of a particular rhetorical stance and style of reviewing which privileged idealised 'intelligent general readers.'[88] Given the anonymity of these reviews, it is difficult, moreover, to make inferences about their authors' horizon of expectation, and the interpretative presuppositions which they brought to their reading of *Influences*.

Trans-Atlantic flows: the international circulation of *Influences*

During the summer of 1910, in correspondence with John Scott Keltie, Semple discussed the desirability of arranging a British edition of her forthcoming book. Following Keltie's advice, Semple 'left the disposition of the English rights' to her publisher Henry Holt, who arranged for the London firm Constable and Company to act as the book's European distributor.[89] Upon the book's publication in the United States, Holt shipped '150 and 10 free copies in sheets' to Constable in London, which were then bound and sold at 18 shillings.[90] Early in 1912, Constable ordered a further 100 copies from Holt – an indication of the book's comparative success.[91] Holt's stock of unbound copies had, however, been exhausted, and they shipped the outstanding order with their own binding.[92] Although it is unclear precisely when Constable offered *Influences* for sale in Britain, the book was reviewed first by *The Bookseller*, an organ of the United Kingdom book trade, on 29 September 1911.

Unlike the early summaries that accompanied *Influences'* publication in the United States, *The Bookseller* offered an extended description of the book's content, method, and intended audience. In a highly complimentary assessment of Semple's work, *The Bookseller* noted that 'the skill with which she marshals her facts and makes her inductions at once arrests and retains the interested attention of the reader.'[93] In common with the *Springfield Daily Republican*, *The Bookseller* drew its readers' attention to the sections of the book most likely to correspond with their personal environmental experiences: 'English people will naturally turn to the chapter describing the main characteristics of island peoples.'[94] *Influences* was, in this way, seen to have a local geographical significance for its British readers. *The Bookseller* also noted that, as a consequence of the 'mass of facts instanced and her comprehensive knowledge of her wide and important subject', Semple's conclusions were of potential interest 'to all races.'[95] *Influences* was important both locally and globally.

The perceived relevance of Semple's book mattered somewhat less to *The Morning Post*, a conservative London daily. Under the headline 'The brotherhood of man', the *Post*'s reviewer explained how *Influences* supported and confirmed a monogenist understanding of human development (the notion which holds that all human races share a single biological origin).[96] Questions of human origin underpinned much eighteenth- and nineteenth-century discussion of biology, anthropology, philosophy, and religion.[97] Although such debates were influenced by theological principles, novel theories of transmutation, evolution, and speciation were significant spurs. These new ideas had an important bearing upon understandings of racial inferiority and superiority. Although the debate was settled, to some degree, during the 1870s – when, along with other works, Darwin's *Descent of Man* (1871) effectively refuted the premise that race was akin to species – the promotion and discussion of monogenist and polygenist perspectives remained live. The persistence of polygenism reflected an unwillingness to concede the troubling moral implications of monogenism: that there existed a 'common ancestry for black and white, Christian and Pagan, cultured and barbarous.'[98] As the *Post*'s reviewer conceded, this was 'a view of humanity not wholly pleasing.'[99]

Readers of the *Post* who had travelled 'off the beaten track' – perhaps 'camping with the lonely Indian on his native lake shores, musing amid the scented turmoil of Eastern bazaars, watching sleek Kanakas fishing in some Queensland lagoon, or swarthy Levantines quarrelling on the quays of Scutari' – could not have failed, its reviewer noted, to 'resist the curious conviction that, after all, there was something in the old Biblical version of a human race dispersed from a common centre and gradually moulded in different patterns by the tyranny of environment.'[100] What Semple's text had done, for the *Post*'s reviewer at least, was to put these suspicions on a scientific footing by dint of its scholarship: 'it is in working out a thousand interesting results that Miss Semple overwhelmingly convinces us.'[101]

It seems unlikely that the reviewer's opinion as to the persuasiveness of Semple's argument would have been shared by the *Post's* editor, Howell Arthur Gwynne (1865–1950).[102] Gwynne held strongly anti-Semitic views, and the *Post* was an occasional organ for these; particularly following the publication in English of the fraudulent *The Protocols of the Elders of Zion* (1920), a text which alleged a Jewish plot to achieve world domination.[103] Part of Gwynne's prejudice depended upon the notion of Jewishness as a racial/species category, and the political perspective of his newspaper reflected this to some extent.[104] Semple's implicit effort to undermine such categorizations would seem, then, to contradict the paper's editorial stance, but did not detract in any overt way from the reviewer's commendatory assessment of *Influences*.

The themes of scholarship, local relevance, and scientific authority were equally apparent in a review published in the *Irish Times*, Ireland's leading unionist newspaper.[105] For the *Times*, Semple's book represented 'one of the most important books ever published upon generalised geography.'[106] The veracity of this claim was demonstrated by reference to Semple's impressive scholarship: by supporting her arguments with 'an infinite variety of instances', she was seen to have produced an 'encyclopædia of geographical facts.'[107] In the scope, ambition, and industry of her work, the *Times* saw fit comparison only with Darwin, but conceded that Semple's text was not 'illustrating anything so wonderful and new' as a theory of evolution by means of natural selection.[108]

In an effort to persuade its readers of the book's local relevance, the *Times* attended to the aspects of *Influences* which dealt particularly with island environments and with Celtic ethnicity. As the reviewer made clear, 'To us who live in the British Islands the chapter on Island Peoples is of deep interest.'[109] Part of this interest lay in the fact that Semple's book seemed to offer an explanation for Celtic religiosity – which was described as the inevitable consequence of life in 'remote, isolated, or mountainous regions.'[110] Perhaps more significantly, the review quoted at length from Semple's discussion of Irish history and her explanation of the nation's comparative domination by England. In Semple's view, although the Irish 'started abreast of the other Northern Celts in nautical efficiency', they experienced an 'arrested development in navigation' from which they did not recover fully.[111] According to Semple's thesis, Great Britain acted as a barrier to the stimulating effect of commercial and cultural exchange with continental Europe and Ireland consequently 'tarried in the tribal stage till after the English conquest.'[112] Semple's conclusion was that Ireland, as a result of excessive isolation, 'failed to learn the salutary lesson of political co-operation and centralisation for defence, such as Scotland learned from England's aggressions, and England from her close Continental neighbour.'[113]

Despite the unflattering nature of Semple's account, it was, for the *Times'* reviewer, proof that 'Ireland suffered from failure, long before English influence could reach her.'[114] Semple's position satisfied, in some ways, the paper's unionist politics. Indeed, the reviewer expressed ironic surprise that, given the negative influence of isolation on the nation's historical development, 'At this moment, in the twentieth century, Ireland is begging for more complete isolation!'[115] Framed in this way, Semple's thesis was co-opted to fulfil a particular political agenda. Although the review of *Influences* did not straightforwardly misrepresent the book's content, it did present it in a manner intended to appeal to the political bias of the paper's readership. This particular staging of *Influences* demonstrated not only its scholarly worth, but also its local political significance.

Although linked by questions of authority, scholarship, and relevance, the various British newspapers which commented upon the publication of *Influences* did so in unique ways. Although it is possible to say that the British (and Irish) press treatment of Semple was broadly approbative, it does not stand as a straightforward surrogate by which the popular British reception of anthropogeography can be described. Here, again, the question is one of scale: what is lost when considering the national response to *Influences* is precisely what the book meant, and why it was welcomed, in London and Dublin. Yet, if we are to think of scale as a social, rather than geographical category, the metropolitan reception of Semple's work can be seen to be inescapably part of its national reception. Semple's book was read and reviewed in simultaneously overlapping and complex circumstances – defined both by local urban conditions and by more general national concerns. It is not possible to speak of the Dublin reading of *Influences*, for example, as a discrete and spatially-bounded phenomenon without consideration being given to contemporaneous national political and religious issues.

Semple's transdisciplinary appeal

In 1922, Semple was invited by the Librarian of the American Geographical Society, John Kirtland Wright (1891–1969), to inscribe the Society's copy of *Influence* with a short postscript setting out the conditions of the book's production and amplifying her thoughts as to its purpose.[116] In her brief correspondence with Wright, Semple affected a relaxed attitude towards the book's critical reception: 'When the book was finally out, I started around the world and did not hear anything about it for eight months. I was content not to.'[117] This nonchalant air was a little disingenuous: Semple's press clipping bureau had kept her up-to-date with newspaper reviews through her journey, and she received 'Several very appreciative letters from both geographers and historians (*mirabile dictu*)' en route.[118] Moreover, Semple was in touch throughout her journey with one newspaper's literary editor: the *Chicago Evening Post*'s Floyd Dell (1887–1969).

Dell joined the conservative *Chicago Evening Post* as editor of its prestigious *Friday Literary Review* supplement in 1911, having previously written for the socialist monthly the *Tri-City Workers' Magazine*.[119] Semple seems first to have contacted Dell in early 1911, whilst living in Chicago, to ask which of the *Post*'s contributors had reviewed 'Mrs. [Alice] Maynell's last volume of essays.'[120] It is not apparent whether Semple thought the review worthy of praise or criticism. Given Dell's socialist and working class background, and the fact that he had an 'unconventional, "feminist" marriage', he seems, for Semple, an unlikely choice of correspondent.[121] When Semple next wrote to Dell, on 18 November 1911, it was by postcard from Singapore where she had paused in her journey from Hong Kong to Sumatra.[122] In her brief note, Semple sent congratulations to Dell's wife, Margery Currey, for an unspecified achievement (most probably in relation to Currey's suffragist work). What Semple did not mention, however, was the *Post*'s review of *Influences*, which had been published the previous week.

Despite Dell's personal connection with Semple, the *Post*'s anonymous review was notably even-handed in its assessment. Unlike several earlier reviews, however, the *Post* assumed of its readers no pre-existing knowledge of Ratzel's work, this despite the fact that the paper was 'directed to the business and professional elite' of Chicago.[123] The review detailed the development of Ratzel's perspective on anthropogeography and described the ways in which Semple had modified them for 'the English reading public.'[124] Semple, on the other hand, was seen to require no introduction: she was 'known to the public for a long time as a contributor to geographical magazines and as author of "American History and Its Geographic Conditions".'[125] For the *Post*'s reviewer, Semple's project of adapting and restating Ratzel's basic principles had been achieved most elegantly by substituting 'facts taken from the American continent for the illustrations given by Ratzel.'[126] In this respect, Semple's incorporation of contemporary scholarship – particularly 'the publications of the Smithsonian Institution' – was 'used to great advantage.'[127]

The work upon which Semple drew was that of the Smithsonian's curator, Otis Tufton Mason (1838–1908). Mason's 1896 monograph *Primitive Travel and Transportation*, which Semple cited at length, was an attempt to understand the historical development of different indigenous populations by reference to their material culture. The influence of environmental circumstances upon this development was, in Mason's view, a significant explanatory factor.[128] His ethnographic examination of the material artefacts of these different populations seemed to lend tacit support to Semple's interpretation of environmental causation. As he noted, 'like [environmental] causes produce like [social] effects.'[129] Although Mason's interpretative stance later changed following criticism by Franz Boas, Semple elected to refer only to those aspects of Mason's earlier work which clearly supported her position.[130] For the *Post*'s reviewer, perhaps unaware of Mason's change of heart, Semple's selective referencing was convincing.

The *Post* was disappointed, however, by the quality of *Influences'* illustrations, which were rather more naive than those which had accompanied Semple's *American History*. As the reviewer noted, 'A work of such tremendous importance ought to be provided with splendid maps and charts, which, unfortunately, are lacking.'[131] Whilst Semple's literary style was not subject to criticism, the accessibility of her language was.[132] For the *Post*, it would have been preferable if 'technical terms used in the book had been translated or explained, for even Webster's Dictionary fails in many instances to give their meaning.'[133] Such limitations did not detract from the book's intrinsic value, and the *Post*'s reviewer concluded that 'Sociologists, anthropologists, economists and geographers will be equally interested in the book, which is an extremely valuable addition to all four of these sciences.'[134]

Quite how significant the book's cross-disciplinary appeal might be was made clear the following week when *The Dial* – a Chicago literary magazine – published a review of *Influences* by a marine biologist, Charles Atwood Kofoid (1865–1947). *The Dial* was not a specialist academic publication, but addressed 'the interested, informed general reader.'[135] Although Kofoid's research interests were principally in relation to oceanic plankton, his intellectual hinterland was broad. He was a 'collector of books' and 'an industrious investigator and reader.'[136] As a consequence of his wide reading, Kofoid contributed 'several thousand reviews' to professional and popular periodicals on various topics relating broadly to biology.[137] It was from a biological vantage point, then, that Kofoid approached Semple's book. For Kofoid, the logic of Semple's argument was not in doubt, but its contemporary applicability was. The industrial and technological developments of the nineteenth and early-twentieth centuries were evidence, Kofoid believed, of the 'elimination of geographic environment as a predominant factor in man's evolution.'[138] As he framed it, the discoveries of modern science were 'fundamentally changing his [man's] relations to the physical configurations of the earth ... and modifying, indeed often minimizing, their effects upon his social and national evolution.'[139] The technologies of mass communication were seen to have circumvented the controlling limitations of geography by facilitating the 'intermingling of the peoples of the earth.'[140] Kofoid believed that Semple's failure to address this fact was a notable weakness.

Semple was praised, however, for presenting Ratzel's work in a 'less dogmatic' fashion, and for providing a thorough analysis of the topographic and climatic factors which related to human development.[141] Where Semple's analysis was lacking, Kofoid felt, was in relation to heredity. In Kofoid's view, heredity was 'a counterfoil ... to the effect of environmental factors.'[142] He saw that in the 'higher levels and later stages of human evolution' the inheritance of genetic material from 'great leaders' was a more powerful influence in shaping the future development of a society than were environmental conditions.[143] Semple had made a conscious choice to concentrate her analysis upon geographical conditions, rather than the 'internal forces of race', in part because she saw the former to have been 'operating strongly and operating persistently' throughout human history.[144] Given that the environment was, in Semple's view, 'a stable force', and one which 'never sleeps', it could be considered 'for all intents and purposes immutable in comparison with the other factor [heredity]' in explaining the historical development of human society.[145] Whilst Semple's book had failed to Kofoid's satisfaction to adequately engage with this important biological principle, he saw value in her work: to 'the biologist and historian', *Influences* was 'of unusual interest.'[146]

The most effusive popular review of *Influences* appeared on 21 December 1911 in *The Nation* – a weekly New York City magazine, and the advertising venue of choice for Semple's publisher, Henry Holt.[147] The anonymous review, dripping with superlatives, was similar in tone to that published earlier by the *Boston Evening Transcript* – that is, Semple's book was presented as a significant scholarly accomplishment and, more importantly, a national triumph. For *The Nation*, *Influences* was 'a remarkable book, one of the few products of American contemporary science which may safely challenge the best that has been put forth in this field by any foreign scientist whatsoever.'[148] Semple's book was seen to have not just a parochial significance to the geographical community, but to have a truly national importance: Semple had demonstrated that, in terms of intellectual achievement and scholarly rigour, the Unites States could equal or exceed any other nation. By so doing, she had also subverted the erroneous conflation of femininity and unreason.[149] As *The Nation* made clear, 'Let us add, without any condescension, that it [*Influences*] places Miss Semple among the handful of women in the world over who are the peers of the foremost men of science.'[150]

For *The Nation*'s reviewer, *Influences* was 'a model of logical arrangement and clear statement', and, in this respect, a significant improvement upon Ratzel's original scheme.[151] The particular value of Semple's approach lay in bringing together 'geography, anthropology, history, and economics' and making clear 'the causal relationships between one and another of these.'[152] By drawing upon these disparate sources, Semple rendered her conclusions into a valid and scientific form that could not 'be gainsaid.'[153] Put simply, she had produced the text that '[Henry] Buckle dreamed of', but had failed to realise.[154] Rather than the monocausal determinism Buckle had advanced, Semple's work was seen to represent a more restrained and considered multicausal perspective.

Making clear the 'two, three, or more causes that contribute to any given effect', indicated the complexity of the relationships with which Semple was dealing, and explained quite why she refrained from 'summing up her immense investigations in the form of a general law.'[155] The reviewer's intention was not to suggest that Semple's book was unscientific or inadequately researched, indeed he or she was unable to recall 'a scientific book which contains more facts on a page than hers.'[156] It was as a scholarly and logical indication of how the subject of anthropogeography might in the future be approached, rather than as a collection of definitive statements about the relationship between humanity and the physical environment, that Semple's book was seen to have most value. By combining unimpeachable scholarship and a style 'enriched by memorable phrases', *Influences* could not, *The Nation* concluded, 'fail to sink deep in many minds.'[157]

News of *The Nation*'s highly laudatory assessment of her book did not take long to reach Semple. She wrote to her editor from Ceylon (Sri Lanka), clearly enthused by the review: 'Did you see the glorified review on my book in the "Nation" of Dec. 21? It makes me eager to get to work again.'[158] Her editor's reply shows that, as far as Henry Holt was concerned, *Influences* was both a critical and commercial success: 'Yes, the Nation review was of quite the right sort, and is of a piece with comments our travellers [sales representatives] are hearing from college people. I think you will be pleased with the report of sales I sent you some weeks ago. The book has done enormously well for the short time it has been before the public, and promises to do better.'[159]

In the popular periodical and newspaper press, *Influences* was subject to varied and distinct interpretations. Opinion differed, for example, as to the book's specific strengths, particular failings, and intended audience. For certain reviewers, Semple's book spoke to very particular and specialized audiences; for others, it had relevance for the general reader (however defined). *The Sun* found *Influences* to be, for example, 'an admirable piece of work, provided it is not used as a text book.'[160] The *Newark News* concluded the opposite: that Semple's book was of singular importance as a 'guide and aid to present-day students.'[161] Opinion varied also as to the scientific character of Semple's geography. For *The Nation*, her text represented the work of a 'true scientist.'[162] For *The Saturday Review* (London), by contrast, there was doubt as to whether Semple's conclusions were 'properly to be considered scientific' at all.[163] Where agreement was near uniform, however, was in relation to Semple's scholarship – to her original observations and her presentation of supporting facts.

Several newspapers, including *The Manchester Guardian*, understood Semple's wish to bring together the perspectives of 'anthropologists, ethnologists, sociologists, and historians' in her exegesis on environmental influence to be its particular strength.[164] It was not simply that Semple's command of these disparate subjects was impressive, but that, in their combined presentation, these different disciplinary positions demonstrated that the 'geographical element has been acting steadily and persistently' in relation to human development.[165] Quite what the professional representatives of these different disciplinary positions made of Semple's engagement with them, became apparent only towards the close of 1911 as professional and scholarly journals delivered their verdict on her book. It is to this question – to what might be termed the professional reception of *Influences* – that I now turn.

RECALLING SCHOLARLY REVIEWING COMMUNITIES

In September 1911, *The Journal of Geography* published the first review of *Influences* from geography's disciplinary perspective.[166] Edited by Richard Dodge, with the intellectual backing of William Morris Davis, the *Journal* had, until 1902, appeared as the *Journal of School Geography*.[167] In both its titular incarnations, the *Journal* was the venue for Semple's first geographical publications, and also carried various articles relating to geography's engagement with environmentalism.[168]

As a supporter of Semple's work, the *Journal* offered a review of *Influences* that was highly complimentary but also even-handed. In an echo of earlier praise for Semple's scholarship and national intellectual contribution, the *Journal*'s reviewer stated 'This volume [*Influences*] is unquestionably the most scholarly contribution to the literature of geography that has yet been produced in America.'[169] The fact that Semple had contributed to the intellectual life of the nation mattered almost as much as did her geographical achievement. As the review's author noted 'that a volume of such evident and unquestionable merit has been produced by an American geographer, is a matter of just pride to us.'[170] In part, Semple's perceived merit was an index of her scholarship, which was considered to be beyond reproach. By making 'nearly one thousand five hundred separate citations of authorities' she had ensured that her text was 'not open to criticism.'[171] Indeed, the *Journal* considered Semple's research to be 'simply prodigious' and concluded that 'for her heroic work the author will receive the unstinting appreciation of geographers and students of geography throughout the English-speaking world.'[172] Where the *Journal* identified potential for censure, however, was in relation to Semple's pronouncements on the geographical background to history. As the reviewer made clear,

> Every thoughtful reader will find here and there that the author [Semple] has drawn conclusions and made interpretations in accordance with preconceived ideas. Being a geographer, and believing in the profound influence of geographic environment, it is not strange if she gives greater weight to the geographical element in history than the ordinary historian would give.[173]

Given the tendency of historians to 'tear to pieces many of the conclusions drawn by other historians', the *Journal* thought it highly likely that 'historians will find some things in the book that they do not accept.'[174]

Whilst Semple's literary style had been identified in popular reviews as an important strength, necessary to the communication of her ideas, the *Journal*'s reviewer considered it to be an impediment to her credibility. Semple's tendency to employ personification was seen to be 'somewhat unfortunate in a scientific treatise.'[175] Although the reviewer recognized that geographers would be able to determine which of Semple's figures of speech were to be taken at face value, it was conceivable that her assertions were 'capable of being taken with many varying degrees of literalness by different readers, and hence leaving different impressions with different readers.'[176] The recognition that, as a result simply of her style, Semple had exposed herself to misinterpretation was significant. There was an important tension between Semple's desire to communicate her anthropogeography using the literary flair which she and Ratzel considered appropriate, and the necessity in the mind of the *Journal*'s reviewer of formulating her ideas in a robust and scientific form. The fact that Semple was unwilling to advance 'hard-and-fast rules' in relation to her anthropogeography was, to an extent, reflected in the construction of her prose.[177] Despite the concerns expressed by the *Journal* in relation to Semple's overuse of personification, it was certain that 'Much that she has set down will stand' and that, consequently, *Influences* was a text for which 'present geographers cannot but feel a deep sense of gratitude.'[178]

The service which *Influences* had rendered the 'organic side of geography' – a contribution which the *Journal* deemed cause for particular gratitude – was highlighted in an anonymous review published in the 23 November 1911 number of *Nature*.[179] As the review's author was keen to point out, geology and mathematics had lent a 'definitiveness and precision to the inorganic side' of geography which was then notably absent in the subject's attention to human social organization.[180] Whilst Ratzel's work had 'furnished a basis for the scientific development of this part of the subject', the absence of an adequate expansion of his perspective, particularly one in English, was seen to have acted as an impediment to the furtherance of human geography.[181] Semple's *Influences* had, for *Nature*'s reviewer, a 'particularly valuable' role in closing this gap.[182] As was noted, 'Precise description and quantitative treatment by recognized scientific method is much needed in this branch of geography, and Miss Semple has placed English-speaking geographers under a deep obligation by her scholarly treatment.'[183]

Nature's reviewer was aware, however, that although Semple's method was scholarly and scientific, it was not strictly nomothetic. Given that Semple's anthropogeography was 'being but gradually evolved', she had intentionally avoided 'Definitions and systematic classification.'[184] Whilst the reviewer regretted that Semple had not been more firm in her convictions and made 'some provisional efforts in this [nomothetic] direction', he or she recognized that the principal value of *Influences* lay in its indicative and suggestive qualities: that it provided a spur to new research and that it was the responsibility of future geographers to 'carry forward the investigation into specific instances in order to determine the value of the different factors involved in each case.'[185] The fact that *Influences* was not a definite explanation of Semple's anthropogeography did not, for *Nature* at least, diminish its intrinsic value. In the reviewer's opinion, Semple's real achievement was in having formulated a rigorous and well-supported methodological framework around which future conclusions could be built, not in providing a set of definitive principles.

Science, geographical methodology, and national pride were themes evident in the laudatory assessment of Semple's book by Ray Hughes Whitbeck (1871–1939), published in the *Bulletin of the American Geographical Society*.[186] Whitbeck, then Associate Professor of Geography at the University of Wisconsin-Madison, was 'an environmental purist to the end.'[187] Educated at Cornell University, where he obtained an A.B. in 1901, Whitbeck had come under the influence there of Ralph Tarr, an important supporter of Semple's early work.[188] Whitbeck developed research interests in various aspects of environmentalism, but was particularly concerned with the 'effects of glaciation on man's activities.'[189] A number of his papers on environmentalist topics appeared in *The Journal of Geography*, of which he was editor between 1910 and 1919. It was under his guidance that the *Journal*'s equitable review of *Influences* had been published.

Whitbeck's editorial balance was apparent also in his review for the *Bulletin*. For him, Semple was seen to embody 'four factors not often within the reach of one person', namely 'deep interest in a great subject, ability to handle it, training, and leisure.'[190] Whitbeck did not intend the last of these qualities, leisure, to appear pejorative; he was eager to make clear that the ambitious scope and scholarship of *Influences* could not have been achieved without 'a prodigious amount of labour' – something which depended upon Semple's part-time teaching responsibilities.[191] In this sense, there could be few other geographers, in Whitbeck's opinion, able and capable of completing the task of adequately reformulating Ratzel's ideas and in so doing of putting geography in the United States more firmly on a scientific footing.

Like *The Journal of Geography* and *Nature*, the *Bulletin* was conscious of the fact that, despite the rigour of Semple's approach, it was not possible for her conclusions to be formulated in a definitive manner. As Whitbeck noted, 'Miss Semple, or anyone else, who attempts to estimate the actual weight of geographic influences in history or development of a people, attempts the impossible.'[192] Whilst Semple's statements were praised for being generally 'conservative and guarded', there were occasions, Whitbeck believed, where Semple's enthusiasm for her thesis was communicated too immoderately: 'there are frequent statements ... which, if taken literally, seem extravagant.'[193] Whitbeck's concern was the same as that of the *Journal*: that Semple's literary style and tendency to use figures of speech could lead to misinterpretation of her work. As he noted, 'a careful and friendly reader can not escape the conviction that the author has aimed to be conservative. An unsympathetic reader may not grant that she had always been successful in this endeavour.'[194]

A further parallel between the *Journal*'s review and that by Whitbeck was in the literal quantification of Semple's scholarship: both reviews noted that her book contained 'nearly 1,500 citations of authorities', and that the chapter on island people alone was 'followed by 233 references.'[195] Semple's citatory tendencies were seen to have a two-fold significance: they acted at once to strengthen her environmentalist claims by situating them within an established and respected literature, and, more pragmatically, to provide 'geographers of the English-speaking world' with an accurate and current bibliography.[196] The authorities upon whom Semple drew spoke not only to her own intellectual concerns, but also to those of her intended audience. Indeed, as one historian of citatory practices has suggested, 'the scholarly authorities whom an author chose to cite must tell us something of their scientific self-fashioning, of their intellectual tastes and imagination, of the sense they had of whom they were in dialogue with as they composed their books.'[197]

The fact that Whitbeck was predisposed towards Semple's anthropogeography can be seen to have conditioned his engagement with *Influences* and his subsequent assessment of it. What is apparent also, however, is that Semple's book was seen by him to represent a particular disciplinary and national achievement. By making a contribution both to geography and to the United States, Semple secured Whitbeck's esteem:

> If the reviewer were disposed to look for faults in the book they doubtless might be found. But the great service which Miss Semple has done for Geography, the years of work which the book has cost, the pardonable pride which we feel in knowing that an American Geographer did the work, all impel this reviewer, at least, to dwell upon the excellencies of the book rather than to seek minor points of weakness.[198]

It did not matter that Semple's conclusions were not definitive or that her prose was bombastic: these small failings were counterbalanced by her more significant contribution to the disciplinary reputation of geography and the United States' intellectual standing.

Despite *The Journal of Geography*'s fear that Semple's ideas were likely to be attacked by academic historians, her book received positive assessment in *The American Historical Review* from Orin Libby, then chair of history at the University of North Dakota.[199] Libby was an enthusiastic disciple of Frederick Turner's frontier thesis, and contributed a number of original papers on the role of the frontier in the historical development of the United States.[200] Semple and Libby had both contributed papers to a session of the American Historical Association chaired by Turner in 1907. The session was intended to bring together geographical, historical, and environmentalist perspectives: the intellectual triumvirate which *Influences* sought to represent. Libby and Semple were, in this respect, part of a broader intellectual project, loosely defined by the attempt to integrate geographical and historical perspectives to more satisfactorily explain the historical settlement, current development, and future potential of the United States.

For Libby, Semple's project – 'carried out with scholarly precision and comprehensive grasp of details' – was not an attempt to prove a direct correlation between environmental circumstances and historical development, but rather was intended to explore 'the complexity of the subject.'[201] Semple's hope in so doing was, in Libby's opinion, to show that 'Man is no longer merely the conqueror of natural environment, nor ... the passive creature of physiographic influences.'[202] By making clear the multifarious and variable relationships between the physical environment and social development, Semple's book was seen to be 'a thoroughly scientific demonstration of the vital relation existing between these two great areas of study [geography and history].'[203] In common with Charles Kofoid writing in *The Dial*, Libby considered Semple's anthropogeography to be most applicable to the early stages of human social development: 'With the fuller development of the social and industrial life, physiography no longer acts as directly or openly; its influence becomes more subtle and hidden.'[204] Although Semple was generally aware of this limitation, there were occasions, Libby believed, when the 'temptation to claim for physiography what clearly belongs to any of a half-dozen forces in society' was clearly too strong for her to resist.[205] Libby went on to detail a number of instances where Semple's interpretation of historical events was, in his professional opinion, dubious. He recognized, however, that Semple's tendency to advance the physiographic component of her thesis was, in part, a consequence of her disciplinary orientation: 'The economist has quite another theory to account for the same phenomena, so has the sociologist.'[206]

Taken together, Libby's criticisms were somewhat minor and he conceded that, given the ambitious scope of Semple's project, it was 'impossible to avoid many seeming misconceptions and errors of fact.'[207] As for Whitbeck, so it was for Libby: these failings were trivial and did not detract from the overall correctitude of Semple's conclusions and the consequent value of her work. As Libby made clear, 'a mere enumeration of these [errors] does not invalidate the genuine claim which the subject of anthropo-geography has.'[208] Semple's book offered, in the *Review*'s opinion, a 'new vantage ground for the study of man.'[209] Rather than 'tear to pieces' Semple's work, as *The Journal of Geography* feared a historian might do, Libby revealed that there was value for historians and geographers in the mutual exchange of interpretative positions – bringing together in cooperation disciplines that 'have suffered from ... separation.'[210] *Influences* was shown to matter almost as much to historians as it did to geographers, and it demonstrated that geography as an intellectual endeavour was more than simply history's handmaiden.

Colonel Close and the challenge of scientific geography

Shortly before Semple embarked upon her eighteen-month post-writing sojourn in 1911, she received a note of congratulations from John Scott Keltie. Keltie expressed his excitement at the imminent publication of *Influences* and assured Semple of his hope that 'we shall be able to have a stunning review of it in the Journal by some competent hand.'[211] The competent hand selected for the task was George Goudie Chisholm (1850–1930), a pioneer of commercial geography in the United Kingdom, and a longstanding correspondent of Semple.[212] Quite by chance, Chisholm's reading of Semple's text (and the context for his review) was influenced by a moment of disciplinary crisis.

On 31 August 1911, Sir Charles Frederick Arden-Close (1865–1952), president of the Geographical Section of the British Association for the Advancement of Science, presented a 'perfectly astounding' paper at the Association's annual meeting in Portsmouth.[213] With vociferousness apparently contrary to his position, Close advanced a damaging critique of disciplinary geography, arguing that it was inconsistent in scope, method, and epistemology.[214] His central claim, based upon an analysis of papers published in *The Geographical Journal* between 1906 and 1910, was that geography was inadequately scientific. He concluded: 'geography ... must prove its independence and value by original, definitive, and, if possible, quantitative research.'[215] Close's criticism of geography was part of a longer-standing doubt as to the discipline's position within the Association.[216] In part, these uncertainties reflected a tension between those aspects of the discipline which were largely descriptive – of which travel and exploration narratives formed a part – and those components (too few, in Close's opinion) which were scientific and explanatory.

Close's address drew an immediate response from a number of British and North American geographers.[217] Through a network of private correspondence, geographers debated the implications of Close's assertion and discussed potential responses. A public reaction to his criticisms was slow, however, to emerge. This was due to the fact that Close's argument could not 'be gainsaid', and that certain members of the geographical establishment were keen to avoid overt displays of division.[218] Hugh Robert Mill (1861–1950), librarian of the Royal Geographical Society, was eager, for example, to mitigate press reporting of Close's remarks, and succeeded in persuading a reporter from *The Times* 'to suppress the controversial part.'[219]

For Chisholm, Close's address afforded an opportunity to articulate and defend geography's intellectual position. Chisholm had been recently appointed lecturer in geography at the newly-established department of geography at the University of Edinburgh, and it was there that he formulated and expressed his initial thoughts on Semple's text.[220] In his opening lecture to the geography class on 11 October 1911, Chisholm, speaking to the title 'Some recent contributions to geography', introduced his students to two newly-published works: Jean Brunhes' *La géographie humaine* (1910), and Semple's *Influences*.[221] For Chisholm, these texts were noteworthy because they could be seen to satisfy Close's opinion that geography should display original, definite, and quantitative research. In addition to its value in countering Close's criticism, *Influences* also complemented Chisholm's belief that 'it is of the highest consequence to have a class of investigators whose constant and single aim is to see that the known causes that affect the value for man of place are never overlooked.'[222] Chisholm's enthusiastic response to *Influences* can be seen both as reaction to its content (which mirrored, to some extent, his own geographical interests), and to the fact that it spoke usefully to a then-current disciplinary debate.

The tone of Chisholm's review, published in the January 1912 number of *The Geographical Journal*, was defined by its introductory sentence: 'There can be little hesitation in pronouncing this the most notable work that has yet appeared in English on the subject to which it is devoted.'[223] In common with certain earlier reviewers, Chisholm found that 'the only English work that can be fairly compared with it [*Influences*]' was Buckle's *History of Civilization in England* (1857–1861).[224] This was not intended to be a backhanded compliment on Chisholm's part – since Semple had pointedly dismissed Buckle's pseudoscience – but rather was recognition of the ambitious scope of her volume. Semple's book was, for Chisholm, a valuable corrective to Buckle's erroneous reasoning: by 'making geography rather than history the foundation of the investigation', Semple had avoided the interpretative limitations evident in Buckle's treatise.[225] It is clear that Chisholm had a certain familiarity with Ratzel's work, and that he assumed the same familiarity in his audience. He recognized, for example, that it was 'to the first of Ratzel's two volumes that Miss Semple's book most closely corresponds.'[226] Whilst he considered Semple's treatment of Ratzel's work to be in most respects superior to the original, he felt that her decision not to include 'the chapter on the vegetable and animal worlds' was an 'important omission.'[227]

Chisholm reassured his readers that 'most of the important ideas' which Ratzel communicated in the second volume of *Anthropogeographie* had been retained by Semple where her scheme permitted. As Chisholm noted, however, certain other components of that volume were not, at the time of their formulation by Ratzel, 'ready for scientific treatment', and Semple's decision to exclude them was justified.[228] One aspect of Ratzel's second volume which Semple, to her disadvantage, did not address was that on 'the structural works of men.'[229] Chisholm felt that an attempt by Semple to attend to this aspect of Ratzel's project and to place it on the same scientific footing as the rest of her volume 'would have been welcome.'[230] Whilst other reviewers of *Influences* merely hinted at their familiarity with Ratzel's principles, Chisholm made his explicit. It was apparent that he had returned to Ratzel's work on several occasions in the preparation of his review, and was able to present a highly detailed comparison of the texts.

In presenting a detailed evaluation of Semple's book, in which he attended to each of Semple's chapters in turn, Chisholm addressed 'the student of geography' and the 'Readers of this *Journal* [who] are already acquainted with Miss Semple' – in short, the professional audience for her book.[231] Although the *Journal*'s review was largely descriptive, rather than evaluative, Chisholm felt justified in suggesting 'one cannot be too emphatic in expressing the value of this work.'[232] The significance of *Influences* was defined by more than its didactic qualities. By acknowledging the complexity of the topic and by presenting the potential scope of future research, Semple had 'left [it] to the student to find out what those [environmental] causes are, and in what manner they have the local attachment indicated.'[233] In this respect, *Influences* was suggestive of how the problems of local environmental influence might be approached. The onus was on geographers working in their own familiar locales to address the detailed and complex interrelations between place and environment. Semple's book provided the methodological framework and intellectual principles, but it was up to the reader, in Chisholm's view, to apply them to their local geographical research.

The value of *Influences* as a refutation of Close's claim that disciplinary geography was unscientific was again alluded to in an anonymous review published in the *Scottish Geographical Magazine*. For the *Magazine*'s reviewer, *Influences* was

> a satisfactory answer – if an answer were required by one of unbiased mind – to the charge lately made that geography is not a science, but a hanger-on of other sciences, a picker-up of crumbs falling from their table, with a suspicion of larceny when unobserved.[234]

The vehement quality of this statement indicates that its author was responding not only to Close's criticism of geography, but also to those expressed more generally by biased commentators in the academy. It is likely that this position also reflected the opinion of the *Magazine*'s editor, Marion Newbigin (1869–1934).[235] Placed in opposition to Close, Semple's book was being co-opted to perform a role for which it had not been designed: defining *and* defending the scope of disciplinary geography.

Whilst the *Scottish Geographical Magazine* and *The Geographical Journal* together saw Semple's book as a timely and welcome contribution to the development of the discipline (particularly in view of the concerns expressed recently by Close), this did not represent a common British response to *Influences*. That Semple's book was understood in other ways – that is, not simply as a response to Close's attack – is apparent in a review by Herbert John Fleure (1877–1969) for *The Geographical Teacher*. Perhaps because Fleure's review was written some time after Close's criticism of university geography, or perhaps because he was addressing an audience of school teachers of geography (for whom debates about the place of geography in the university were of less immediate concern), Fleure chose not to situate his assessment of *Influences* in the way Chisholm and the *Magazine* together had done.

Fleure was Professor of Zoology at the University College of Wales, Aberystwyth, where he specialized in the study of natural regions and human evolution.[236] His geographical interests were closely allied with the regional geography of Paul Vidal de la Blache and Andrew Herbertson, and his perspective on human biological and societal organization drew from Spencer, Darwin, Huxley, and Patrick Geddes (1854–1932).[237] An enthusiastic physical anthropologist and archaeologist, Fleure undertook extensive fieldwork in Wales, describing and classifying racial types. Charting regional variations in language and physical characteristics – the consequence of an interplay between heredity and environment – Fleure demonstrated his possibilist inclination; he showed that 'the modern environment was an end product deriving its character from the activities of settlers over thousands of years.'[238]

As a consequence of Fleure's intellectual orientation, his reaction to *Influences* was 'rather hostile'; he objected to its apparently deterministic environmentalism.[239] As he noted later, 'I have thought of men and environment as knit together – neither dominating the other – and I feel that we lose a lot when we say that such & such a fact is due to environmental influence.'[240] Fleure's review considered the practical application of Semple's anthropogeography, and found her causal description of human–environment relations wanting in several respects.[241] With a nod towards Semple's deterministic rendition of anthropogeography, Fleure found it notable that she did not advocate Henri Bergson's (1859–1941) notion of élan vital as an explanatory cause. Bergson's ideas formed part of a wider doctrine of vitalism, and sought to attribute to evolution a spiritual, non-mechanical guidance – a vital spark that directed the course of evolutionary development. Despite his sardonic aside, Fleure's criticisms of Semple's approach were utilitarian. He questioned the extent to which her generalizations might usefully be applied to the study of local regions, concluding: 'The reader, who tries to apply Miss Semple's theses to the ... study of his own locality ... will find the need of modification in many points.'[242]

Fleure did not read *Influences* as the nomothetic manifesto which Chisholm and the *Scottish Geographical Magazine* had identified. Distanced sufficiently from Close's climacteric address, and influenced by his own perspective on regional geography, his reading was rather more critical and considered. Fleure did not reject Semple's claims entirely, however, and employed *Influences* in a pedagogic capacity at Aberystwyth, just as it was at other British academic institutions.

Such scholarly engagements provide an important means by which to recover the different ways in which Semple's ideas were conveyed to students – as a teaching resource, as a methodological guide, as an indication of what geography should or should not be. They also demonstrate that the reception of *Influences* was not a temporally-fixed event, but was a continual process of negotiation. Before moving on to examine these scholarly engagements, I should like to consider further what *Influences* meant to its non-geographical reviewers (social scientists, economists, and political scientists), and to geographers working outside the Anglophone context.

Non-geographers reading

That Semple's book was potentially of interest to more than its geographical audience was made evident in a review by the economist George Byron Roorbach (1879–1934) published in *The Annals of the American Academy of Political and Social Science*.[243] Roorbach's opinion, which mirrored closely that of Chisholm, was that Semple's book 'must be regarded as the most valuable contribution to the subject of anthropo-geography that has yet been published.'[244] Beyond *Influences'* obvious appeal in this respect to geographers, Roorbach considered it to be 'of great value ... [to] the student of the social and political sciences, and of absorbing interest to the intelligent general reader.'[245] Roorbach, who shared Chisholm's research interests in commercial and economic geography, was aware that, in addition to being a useful restatement of Ratzel's principles, Semple's book also had a wider political significance for contemporary geography.[246] *Influences* was 'a good illustration of the meaning and value of scientific geography.'[247]

On 15 December 1911, Davis Rich Dewey (1858–1942), editor of the *American Economic Review*, wrote to the University of Minnesota economist Edward Van Dyke Robinson (1867–1915), inviting him to review Semple's *Influences*. Dewey's only stipulation by way of editorial guidance was that 'The review should not run over 700 words.'[248] Beyond this simple constraint, Robinson was free to determine the review's content, scope, and purpose. Although Dewey had offered to send Robinson a copy of *Influences*, this was not necessary since

Robinson was among those scholars to whom Henry Holt sent a copy of *Influences* upon its publication. The list of those who received a copy was compiled by Semple, and included 'those who have been using my previous book, for several years past, as text or reference.'[249] It is unclear quite how Semple and Robinson became acquainted, but it seems likely that as a frequent contributor to *The Journal of Geography*, Semple would have been aware of his work.[250] Robinson had also undertaken a Ph.D. at the University of Leipzig at a time which coincided with Semple's period of study there. Whether or not they knew one another then is uncertain, but it seems likely that their shared experience of Leipzig would have been an important basis to conversation and reminiscence.[251]

As Whitbeck had done in the *Bulletin of the American Geographical Society*, Robinson emphasized the important service which Semple's book had rendered contemporary American science: *Influences* was, in his view, 'on a par with the best in either German or French.'[252] By combining 'geography, anthropology, history and *economics*', Semple had produced 'a truly monumental work which no serious student of any of the social sciences can afford to ignore.'[253] Robinson considered Semple's method to be 'thoroughly scientific' and was unable to detect any instance 'of forcing the facts to fit any prearranged scheme.'[254] Unlike the anonymous reviewer in *The Journal of Geography*, Robinson found Semple's prose style well tailored to her subject: 'the style is always clear, lively and sometimes poetic. As a result, there is hardly a dull page in the book.'[255] Conscious of the audience to whom he was addressing his review, Robinson paid particular attention to Semple's discussion of economic matters. Despite the 'immense literature' from which Semple had drawn, Robinson felt that 'disproportionate use has been made of geography and anthropology, compared to history and economics.'[256] Despite being concerned that 'it may seem ungracious to ask for more', he believed that Semple's reliance on Thomas Malthus and Wilhelm Roscher for economic principles, and George Grote and Quintus Curtius, 'both long since out of date', for Greek history, was rather inadequate.[257] Robinson went on to list fourteen scholars – geographers, economists, and historians – from whose uncited perspective he felt *Influences* might have benefited. Although he recognized that attention to these works might not have changed Semple's conclusions materially, he felt that they 'would have immensely strengthened the authority of the work.'[258]

In contrast to the generally mild criticisms which Semple's text had attracted from geographers and economists, it was subject to some fairly robust censure by the Columbia University sociologist Alvan Alonzo Tenney (1876–1937) in a review published in *Political Science Quarterly*.[259] Tenney was a social theorist, with research interests in 'population, public opinion, and international peace.'[260] In his 1908 volume *Social Democracy and Population*, Tenney 'attacked "anthropo-sociologists" and other biological determinists', arguing that 'intelligent knowledge of biology allowed increased "social democracy".'[261] In short, Tenney's belief was that knowledge, whether scientific or not, had a greater role in shaping society than did biological or environmental factors. In this respect, his position was not far removed from those critics who considered modern scientific developments to have circumvented the role of environmental influence.

In discussing Semple's method – that of comparing 'typical peoples of all races … living under similar geographic conditions' in order to show that 'similar or related social, economic or historical development' was a consequence of environment rather than race – Tenney identified a 'serious theoretical and practical fallacy.'[262] As he made clear, 'Unless undue extension is given the terms race and geographic environment, Miss Semple … has taken no account of a very important third factor, namely, knowledge.'[263] In Tenney's view, social, economic, and historical similarities were more likely to be the result of 'cultural contacts and the spread of institutions by imitation' than of geographical factors in isolation.[264] Tenney was concerned, then, that 'the unwary reader may often fail to appreciate the importance of non-geographic factors not mentioned in the text.'[265]

In addition to the potential speciousness of Semple's conclusions, Tenney thought she was inconsistent in her claim that geographical conditions were immutable in their influence. He identified a number of occasions where Semple contradicted her position by acknowledging the variability of one or another geographical factor. At the same time, however, Tenney felt Semple had also failed to address the long-term variations in climate and topography that would have influenced the development of 'primitive man.'[266] Semple's failing, essentially, was a lack of adequate temporal appreciation: she was guilty of not thinking 'in tens of centuries.'[267] As a consequence of these apparent failings, Tenney considered *Influences* to have limited nomothetic value:

> The reader who expects to find in the volume a succinct and coördinated statement of principles and a well-constructed theory in which there is adequate presentation of the importance of the various geographic influences on man in relation to each other, together with their importance as a whole in relation to other influences, will be disappointed.[268]

Despite this fairly damning indictment, he conceded that *Influences* was 'remarkably well written', and, although he was intellectually opposed to Semple's project, he felt it was 'a work which no student in any branch of political science can afford to overlook.'[269]

Beyond the Anglo-American world

Although Semple's intention had been to adapt Ratzel's work to the 'Anglo-Celtic and especially to the Anglo-American mind', it is apparent that its impact and readership were rather more international.[270] Two foreign-language reviews of *Influences* – one Italian, one German – provide an interesting indication of how Semple's work was engaged with in these distinct cultural and intellectual contexts. Whilst, as has already been elaborated, these reviews cannot be seen to represent straightforwardly the Italian or German reading of Semple's book, they serve as an important counterpoint to its predominantly Anglo-American interpretation, and reveal the commonalities and differences in styles of reviewing practice.

It was to the newly-appointed lecturer in geography at the University of Padua, Roberto Almagià (1884–1962), that the *Bollettino della Società Geographica Italiana* turned in 1912 for its review of *Influences*. Almagià, who became one of Italy's most distinguished geographical scholars, had studied under the discipline's modern founder in that country, Giuseppe Dalla Vedova (1834–1919).[271] Vedova had been an enthusiastic proponent of 'the modern methods of geographical study which had already borne fruit in Germany', and Almagià's exposure to, and familiarity with, these works, particularly in relation to Ratzel's geography, is apparent in his review of Semple's book.[272] Like Chisholm in *The Geographical Journal*, Almagià sought to provide a detailed comparison between Semple's book and Ratzel's original text. Almagià was generally satisfied with Semple's interpretation of Ratzel's themes, but regretted, as had Chisholm, the fact that *Influences* lacked 'a systematic exposition of the influences of the biological environment on man.'[273] Despite this empirical omission, Almagià felt that Semple's project had succeeded in placing Ratzel's principles 'in clear light' by supporting them 'with frequent references and numerous examples.'[274]

What set Semple's work apart from that of Ratzel was her 'determined exclusions of bare definitions and theoretical formulae', and the 'abundance of examples ... historical proofs ... [and] factual information' with which she supported her assertions.[275] By making clear the complexity of the environmentalist principles with which she was dealing, and by placing them on a scientific and well-supported basis, Semple had rendered 'a book very apt to introduce into schools' – a fact enhanced by 'the lucidity of the exposition and the easiness of reading.'[276]

Almagià, echoing Chisholm's review, presented a detailed chapter-by-chapter analysis of *Influences*, in which he found Semple's book to be characterized by 'lucidity and orderliness of exposition' and 'accurateness in the research and in the choice ... of [illustrative] examples.'[277] Like Robinson, however, Almagià expressed regret that Semple's selection of authoritative literature had failed to extend beyond the Anglophone. Whilst he acknowledged that Semple had given some consideration to works in German and French, it was clear to him that Semple was 'evidently ignorant of our language.'[278] Semple's inadequate attention to Italian sources was only part of her failing. As Almagià made clear 'we Italians might also complain that not all the observations ... [made about Italy] appear equally exact.'[279] For Almagià, it was inconceivable that

> one might wholeheartedly welcome all that the author says about the consequences of the position of Italy in the Mediterranean ... or about the contrasts between continental and peninsular Italy ... nor might we subscribe to the judgement that the Italian state has

renounced every territorial expansion and accepted its present borders as definitive due to a lack of energy and national purpose![280]

Despite his mild apoplexy at these national slights, Almagià did not feel that they diminished 'the general value of the work.'[281] In the absence of a similarly comprehensive and comprehensible text in Italian on the principles of anthropogeography, Almagià felt that Semple's book would 'be greeted favourably by [Italian] scholars.'[282]

The validity of the *Boston Herald*'s ironic suggestion that Semple's book could 'be advantageously re-translated for the use of Germans themselves' was confirmed by the German geographer Otto Schlüter (1872–1959), in a review published in *Petermanns Geographische Mitteilungen*.[283] Schlüter had been educated under Ferdinand von Richthofen, and had inherited from him (and, by implication, from Carl Ritter also) a desire to focus on the comparative analysis of landscapes.[284] For Schlüter, the physical landscape was a cultural product as much as it was the consequence of a series of natural conditions. In this respect, his outlook differed from the predominately environmentalist perspective which had underpinned much earlier work in German geography. His examination of settlement patterns in the Unstrut Valley – a relatively homogeneous riparian environment – showed how populations from distinct cultural backgrounds used and altered the landscape in notably different ways. In Schlüter's view, to understand these societies, it was necessary to understand their cultural landscape, not simply their physical setting.

Despite having shown through work in the field that a uniform physical environment could support multiple social and cultural expressions – a situation at odds with Semple's anthropogeography – Schlüter seems not to have considered this an impediment to recommending *Influences* to the readers of *Petermanns*. Schlüter commended Semple for having adapted Ratzel's work 'tactfully and with scientific aplomb.'[285] By remaining faithful to the spirit of the original text, but providing a more robust scientific formulation, Semple had produced a 'whole new masterpiece.'[286] For Schlüter, Semple's main achievement lay in providing a counterpoint to 'Ratzel's erratic thought process' by clarifying the central tenets of anthropogeography and by supporting them with systematic reference to examples from fieldwork and contemporary geographical literature.[287] Semple's text would, in Schlüter's opinion, be 'very much welcomed by German geographers.'[288]

Schlüter's disenchantment with the environmentalist tenet came, as it had done for Franz Boas, in the field: at the moment when his observations of societal organization and cultural expression appeared to contradict assumed principles of environmental control. It is unclear, then, why Schlüter was so unequivocal in his praise for Semple's *Influences*. It is probable that he appreciated Semple's desire for scientific rigour and her reluctance to make definitive claims based upon the principles to which she subscribed. The fact that Semple's approach was largely comparative perhaps also satisfied Schlüter's wish to replicate the methods set forth by Ritter and Richthofen. What is clear is that there was no straightforward connection between Schlüter's research concerns and his analysis of Semple's book. Whether his review of *Influences* was in some way disingenuous, or was tailored to fit an unknown editorial position at *Petermanns*, cannot necessarily be resolved. If we are to take Schlüter's review as a genuine reflection of his opinion, however, it would seem to indicate that *Influences* had value beyond its attempt to prove anthropogeography. That Semple's project mirrored Schlüter's wish to place geographical research on a scientific basis was reason enough, it seems, to recommend it to the readers of *Petermanns*.

CONCLUSION: SCALE, INTERPRETATIVE COMMUNITIES, AND THE PROBLEM OF ANALYSIS

In the two years following its publication in May 1911, *Influences* was reviewed in more than forty periodicals, including local and national newspapers; geographical and non-geographical journals; and popular and literary magazines. The diverse character of these publications was matched in variety by the tone of their reviews. Given such diversity, any attempt to identify representative and common themes – to describe particular styles and cultures of reviewing – can only ever be partial. Nevertheless, it is possible to make certain claims about the reviewing of Semple's book which help define and explain the initial character of its reception. Further, the complicated and multifaceted character of these reviews prompt important consideration of the appropriateness of scale (whether social or geographical) as an analytical mode.

Were cultures of reading and reviewing to follow neat geographical scales, then we should be able to describe clear differences between the reading of *Influences* in, say, Boston and New York City, and between Britain and the United States. The fact that, as these reviews demonstrate, there was not straightforwardly a Boston or New York City or British or United States reading of Semple's book, makes clear that the contours of reviewing style do not necessarily follow those of taken-for-granted geographical scales – the city, the region, the nation, and so on. It is apparent that in different places, however, different types of reading were possible. Whilst location did not always *determine* how Semple's book was read, it did *facilitate* certain types of engagement. Scale is not invalid as an analytical category, but alone it is insufficient. In the same way that it is problematic to point to particular metropolitan, regional, or national readings of *Influences*, so too it is imprudent to speak in terms of disciplinary-specific readings of Semple, or of styles of reviewing that were unique to, or characteristic of, particular media. The danger of making generalizations about the types of reading and styles of reviewing is, in part, that the role of individual authorship, editorial policy, and intended audience is diminished. There are, however, some important commonalities that persist and are worth emphasizing.

The most striking of these is that the published reviews, in almost all cases, devoted little effort to explaining what anthropogeography was, or where its intellectual origins lay. It was assumed almost universally by the book's reviewers – whether genuinely, or as the consequence of disingenuous intellectual affectation – that their readers were aware of the work of Ratzel and Semple, and that little additional explanation was required as to the nature and purpose of anthropogeography. Whilst this might have been valid in relation to the readers of geographical publications, it seems improbable that the less-specialized audience of metropolitan newspapers would have been cognisant of Ratzel's ideas. The tendency of reviewers to attribute to their readers this level of intellectual sophistication served, however, a dual function: it spared the reviewer the task of explaining the complex intellectual underpinnings of Ratzel's work, whilst serving also to define the periodical's audience as intelligent general readers.

Stylistic differences between periodicals were somewhat more apparent when it came to describing the content of Semple's book. In general terms, professional periodicals, particularly geographical journals, systematically outlined the content of *Influences* – typically offering a chapter-by-chapter summary. Non-professional reviews, by contrast, frequently presented generalized overviews, occasionally highlighting a specific aspect of Semple's book which would correspond to the local geographical knowledge of their audience. Revealing the local relevance of Semple's book did not, though, necessarily equate to a local reading of *Influences*. The identification of the text's local relevance was a discursive or rhetorical element common to several reviews – it was not uniquely local. Whilst it is true that what precisely was deemed local and relevant varied between periodicals, the very act of identifying the local importance of *Influences* was commonplace.

Where it *is* possible to see a local reading of *Influences*, however, is in the *Irish Times'* review. By making the link between Semple's anthropogeographical analysis of Irish history and the then-current local (and national) concern of Irish independence, the *Times* offered a reading that was situated geographically and politically as well as temporally. The particular political climate of Dublin at the time of the review's publication facilitated a specific engagement with Semple's text – one which, arguably, would not have occurred in quite the same way at a different time or in a different location. We can perhaps make a distinction, then, between local readings (the *Irish Times* review being one example) and localized readings – those which drew attention to the relevance of Semple's book to local contexts, but were not explicitly defined by these contexts.

Similarly, it is useful to distinguish between discipline-specific readings of Semple's work, and more general disciplinary engagements. The former are those reviews which were overtly shaped by the disciplinary context of their author. In this respect, the *Scottish Geographical Magazine*'s review and that written by George Chisholm are important examples – the context and focus of each review was directed in some way by Close's negative assessment of disciplinary geography. Again, this is not to suggest that these were necessarily geographical readings of Semple, but to argue that they reflected a particular disciplinary moment (which, in this case, also happened to be a national concern). In contrast to these readings, where topical disciplinary debates played an obvious role in shaping the assessment of Semple's text, a number of professional reviews considered her text only in abstract relation to their disciplinary concerns. This is particularly evident in relation to those periodicals and authors who listed texts and sources (most related to their own discipline) which Semple had failed to mention. Expressing the wish that Semple had attended more particular to economic texts, as Edward Robinson had done – or that she might have benefited from an engagement with Italian geographical literature, as Roberto Almagià suggested – was to show what it was to read Semple's book in *relation* to a disciplinary context, rather than *through* that context. This is a subtle but important distinction. Robinson and Almagià each assessed *Influences* as it spoke to their professional and scholarly position, but their reviews were not necessarily uniquely shaped by that position.

The reaction to Semple's scholarship – to her extensive citation of authorities – was almost unanimously commended (even in those cases where reviewers identified sources from which Semple had failed to draw). Almost universally, Semple's effort to situate her work within an established literature, and to support her claims by reference to contemporary research, was seen as a warrant of credibility. Securing authority in this way mattered particularly to geographical reviewers of *Influences*, for whom it related to the then-current project of defining geography as an independent and scientific discipline. That geography could be considered a science was a 'strange assertion', particularly for those non-geographers for whom the subject inevitably recalled 'certain grammar-school exercises in locating rivers, mountains, political boundaries ... and in memorizing lists of exports and imports.'[289] Semple's book was seen to have a unique and particular importance in helping to place geography on a nomothetic footing by showing that it was more than simply 'descriptive and mnemonic.'[290]

Beyond the book's immediate relevance to the promotion of disciplinary geography, it was also, for a number of American reviewers, an important national triumph. In some respects, this reading of *Influences* as a distinct contribution to American intellectual life was nationally-specific – the reading was uniquely American. This is not to suggest that it was a position shared by all (or, indeed, most) of the book's reviews in the United States. Rather, we can think of a national reading of Semple's book that was not shared and universal. Whilst the patriotic assessments of *The Nation*, *The Journal of Geography*, and *The American Economic Review* were uniquely American, they cannot necessarily stand as proxies for the national response to *Influences*. I would like to distinguish, then, between the national as a geographical scale and the national as a common social category. The reviews of *Influences* published in the United States varied considerably in terms of style, content, purpose, and assessment. As a result, attempting to identify a common response to Semple's book is potentially specious. It is possible, however, to recognize certain themes which were unique to the United States (principally the book's national contribution). In this sense, whilst these reviews were not representative of a nationwide response (geographically speaking), they were a uniquely American interpretation. We can speak of a national response to Semple's book only if we are willing to see the nation as a social entity, not as a fixed spatial category. In this way, it is not necessary to choose the correct scale of analysis in studying reception, but to explore and justify what we take scale to be and to see how our categories can better make sense of these interpretations.

Whilst the discussion in this chapter of the published reviews of *Influences* has addressed the initial engagement with *Influences*, Chapter 4 and 5 consider its career more broadly. Before detailing individual stories of *Influence*'s reading in different disciplinary settings (as I do in Chapter 5), I consider the influence of Semple's oration – of her public lectures and scholarly seminars – upon the acceptance of her anthropogeography. My attention is to the different 'spaces of speech' in which Semple's ideas were promulgated, discussed, and disputed.[291] She addressed audiences who, for the most part, had not read *Influences*, and might never do so. In this context, speech and public demonstration were central to the diffusion of her knowledge. The reception of *Influences* will, in the chapter which follows, be shown to depend, then, not simply upon its textual content, but also upon Semple's representation of it. Despite its ephemeral character, talk will be shown to have mattered.

4 From the field to the lecture theatre: proving and disseminating anthropogeography

As Semple prepared in 1911 to depart on her eighteen-month so-journ, she received an invitation from Andrew Herbertson, director of the University of Oxford's School of Geography, to lecture there on her anthropogeographical work. Herbertson had followed Semple's progress on *Influences* with keen interest, and was eager for her to articulate her perspective on geography, both in its textual guise and in the form of lectures to his students at Oxford.[1] Semple's other principal British correspondents – George Chisholm and John Scott Keltie – were similarly anxious to persuade Semple to lecture to the societies they represented (the Royal Scottish Geographical Society and the Royal Geographical Society respectively).[2] Their enthusiasm reflected the relevance and timeliness of Semple's anthropogeographical contribution, but also had to do with the potential popular appeal of a narrative based upon her world travels.

Part of the purpose of Semple's journey was to gather additional proofs of her anthropogeography – to show that in different environments her ideas remained valid. Given that much of *Influences* had been based upon an analysis of secondary literature, Semple was eager to test and refine her concepts in the field. Although she saw her journey as an important opportunity to relax after the long effort her book had required, its scholarly purpose was clear: 'I long to see and live anthropo-geographically after theorizing about it for the past seven years.'[3] This chapter considers, then, the connections between Semple's field research and the communication of her anthropogeography. It is concerned with the various venues – public and professional – in which Semple sought to disseminate her work, and attends to the different ways in which she employed the knowledge and experience gained during her excursion. Rather than dealing with the reading of *Influences per se*, this chapter describes a series of communicative moments and speech acts which shaped the book's reception. In this respect, orality and aurality will be shown to be as significant as textuality when accounting for the response which *Influences* engendered.

Studies of science as a form of communication, at least in their consideration of the nineteenth and twentieth centuries, have tended to privilege print as the principal medium for knowledge exchange and dissemination. For one historian of science, the 'apotheosis of print in the nineteenth century has led other forms of communication to seem feeble and ephemeral.'[4] Yet, whilst print has 'dominated ideas of what it means to make a contribution to knowledge', attention to different venues and modes of scientific communication – museums and exhibitions, scientific societies and public lectures – has disrupted the assumed primacy of print.[5] Among these various non-textual modes of scientific communication, the popular scientific lecture has been shown to be important.[6] As a venue for the dissemination of knowledge, the popular lecture was often implicated, through the activities of local scientific and literary societies, in the process of defining civic culture.[7] The provincial lecture theatre was a space of instruction and edification, of debate and discussion – a venue in which rhetorical aptitude and correct social deportment mattered to the successful communication of knowledge.[8] Such 'spaces of speech' were fundamental expressions of Victorian and Edwardian scientific culture, and their contribution to the public understanding of science was comparable to that of the printed text.[9] As the lawyer-turned-geologist Charles Lyell noted in the 1840s, 'The invention of printing, followed by the rapid and general dispersal of the cheap daily newspaper ... have been by no means permitted to supersede the instrumentality of oral teaching, and the powerful sympathy and excitement created by congregated numbers.'[10]

In showing how speech mattered to the diffusion and reception of Semple's ideas, I begin by examining her contribution to the geographical summer school at the University of Oxford, before going on to describe the importance of her lecture tour of Scotland under the auspices of the Royal Scottish Geographical Society. Semple's address to the Royal Geographical Society in London in 1912 will be shown to be significant in conferring upon her a warrant of academic credibility, and in providing a prompt to the admission of women fellows to the Society. I conclude by considering how Semple promoted *Influences* (and the ideas it contained) at various universities and teacher training colleges in the United States. I am interested not only in Semple's pedagogical approach, but also in the ways in which ideas of environmentalism conditioned the development of Anglo-American geography curricula and, thus, the meaning of *Influences* in sites of geographical instruction.

SEEKING PROOF IN THE FIELD

Semple's sojourn began with a visit to Japan was facilitated by two former Vassar classmates – Stematz Yamakawa (the first Japanese woman to receive a college degree, and wife of Field-Marshal Prince Oyama who was commander-in-chief in Manchuria during the Russo-Japanese war of 1904–1905), and Baroness Uriu (whose husband, Rear Admiral Uriu, served in the Imperial Japanese Navy during the war). As a consequence of these personal connections, Semple was able to travel freely and was provided with skilled government interpreters. She spent three months 'studying the geographic factors in the utilization of material resources' in Hondo (now Honshū), and, during a 175-mile journey by foot through the island's central mountain range (which recalled earlier work in the Kentucky Mountains), examined the influence of altitude upon agricultural patterns.[11] She saw the latter as 'the result of climate relief.'[12] Although her exploration of Japan was atypical for a Western female of the period, it was not unprecedented. More than twenty years previously, the English traveller Isabella Bird (1831–1904) had completed an extended exploration of Yezo (Hokkaidō), Japan's northernmost island, before travelling widely

94

in Southeast Asia.[13] Bird's experiences formed the basis of her 1880 volume *Unbeaten Tracks in Japan*.[14] Whilst it was not Semple's intention to replicate Bird's journey, it is apparent that she was familiar with Bird's writings and that these served as useful preparation for her own explorations.[15] Although Bird's observations were not necessarily framed as scientific – at least as the term was understood at the time – her systematic approach was something which Semple sought to replicate.[16] Semple's eagerness to ensure the rigour of her work was clear. She made a specific point after completing her principal fieldwork of visiting the Agricultural College of Tokyo Imperial University to 'check off my own observations against the statistical [accounts].'[17]

Semple's exploration of Hondo's mountain region had an emotional as well as scholarly significance. Perhaps as a consequence of her early fieldwork in the Kentucky Mountains, and her long period of writing in the Catskills, Semple had a particular fondness for, and intellectual interest in, mountain environments. Semple, like Bird, 'revelled in the "glorious upper world"' which the mountains represented, and seemed to relish the physical extremes which she encountered.[18] In a letter to Keltie she recalled, 'I have been alternately blistered by the sun, and stormbound by the typhoon in some mountain village; but it has all been one prolonged delight.'[19] At various stages during her world travels Semple sought out similarly mountainous regions, allowing her to perform a comparative analysis of different elevated environments. In subsequent months, Semple undertook 'walking trips for like purposes through the Hartz Mountains, the Thuringian Forest, the mountains of Norway and Sweden, and through the Alps of Austria and Switzerland and in eastern France.'[20] Whilst these mountain environments held an undeniable frisson of danger and sublimity, Semple's attraction to them was somewhat more pragmatic: they represented an 'anthropogeographical laboratory' in which her principles could be tested and refined.[21]

Following her period of exploration in Japan, Semple passed through Korea and Manchuria, where she was given 'special privileges and passes over all the principal roads, and the best Government guides and interpreters.'[22] A guest of the South Manchurian Railway Company, Semple toured the industrial complexes at Port Arthur (Lüshun), and then travelled to Peking (Beijing) where she explored the Forbidden City. From Peking, Semple journeyed by the newly-completed railroad to the city of Kalgan (Zhangjiakou), an important trading point bordering the Gobi Desert.[23] Semple was hosted there by Anglo-American tobacco industrialists (a number of whom were from Kentucky), who were successfully increasing the market for cigarettes in an area where opium had been recently outlawed. There, Semple undertook a study of desert trade patterns, particularly the camel caravans which ran into Tibet and Gansu.[24]

Semple concluded the Asian portion of her journey with visits to Java, the Malay Peninsula, Burma (Myanmar), India, and Ceylon (Sri Lanka). She devoted considerable attention to Java and Ceylon as island environments, and completed detailed studies of their geographical and social organization.[25] Throughout her journey, Semple was in regular correspondence with Keltie who was eager to ensure that she would 'come to us before ... any other Society in our country.'[26] Keltie's concern that Semple might decide to lecture first to the Royal Scottish Geographical Society was unfounded. Semple was keen to reassure Keltie that 'I should wish to give your Society the preference as to the date for my lecture, as the invitation came first from you.'[27] Moreover, she left it to Keltie and Chisholm to 'arrange the dates between you.'[28] One thing that Semple did wish to make clear, however, was her desire to *present* her paper, rather than simply *read* it. As she noted 'I would rather talk off this lecture than read it from notes. When I get up a good head of steam, so to speak, I can then make the subject more alive.'[29] In the same way that she considered literary prose necessary for the textual communication of her anthropogeography, so Semple also saw the correct performance of her ideas as crucial to their effective transmission. Semple was keen to employ the oratorical techniques she had perfected at Chicago, and feared that having to read from a written text would diminish the impact of her work. Keltie was happy to assure Semple that 'we much prefer that anyone lecturing to us should speak and not read.'[30]

From the Indian subcontinent, Semple sailed to the Mediterranean, where she spent some weeks visiting important centres of ancient Greece and Rome. Semple devoted an extended period to the study of Mediterranean agricultural practices, stock-raising, and, perhaps as a consequence of her recent examination of contemporary trade patterns in the Gobi Desert, ancient trading routes.[31] Again, Semple's wish was to test and to refine her anthropogeographical ideas by applying them in the field. Her investigations marked the beginning of a third distinct phase of her academic research, and were the foundation upon which her final book *The Geography of the Mediterranean Region* (1931) was based. Semple completed her global odyssey with a northward sweep, taking in Switzerland, Germany (where she explored the Thuringian Forest and the Hartz Mountains), France, the Netherlands, Belgium, Sweden, and Norway, before arriving in the United Kingdom in July 1912.

'Listening to her quiet voice': Oxford's geographical summer school

Upon her arrival in England, Semple's initial destination was the Lyceum Club in Piccadilly, of which she was a corresponding member.[32] Organized in 1904 as a public meeting venue for women engaged in literary, artistic, and scientific pursuits, the Lyceum was the first women's club in central London.[33] Using the Club as a social and academic base, Semple undertook research at the library of the Royal Geographical Society and made final preparations for her planned lectures at the University of Oxford. In developing a five-lecture course on 'Island People', Semple drew upon her recent research in Japan and South Asia. Her seminar was intended to include a detailed anthropogeographical analysis of Sicily, Ceylon, Java, Japan, and Great Britain 'as types of island environment.'[34] Semple's plan was to prepare a course which would convey her anthropogeographical ideas through the discussion of contemporary geographical research. Having proved her ideas in the field, she sought to demonstrate them in the classroom.

The Oxford biennial summer schools in geography had been initiated in 1902 by Halford Mackinder (1861–1947) as a forum in which school teachers of geography could extend their knowledge and practical experience of the subject.[35] Although the first meeting attracted only thirty participants, the summer schools went on to exert a significant influence upon the nature and practice of geography education in Britain during the first quarter of the twentieth century. In addition to benefiting from the teaching services of important British scholars, the schools also attracted 'many of the leading contemporary American geographers.'[36] Besides Semple, contributions were made by William Morris Davis and Albert Brigham.[37]

During the summer schools' initial years, much of the teaching load was assumed by Andrew Herbertson, and overall organizational responsibility passed to him, and his wife Dorothy, when Mackinder left Oxford in 1905. The five summer meetings organized by Herbertson between 1908 and 1914 were attended by more than 850 teachers of school geography.[38] Although the meetings' form did not alter radically from that developed by Mackinder, Herbertson tailored them more closely to those summer schools pioneered by Patrick Geddes in Edinburgh during the 1880s.[39] Herbertson's vision was for a course that ranged from the classroom to the field, and embraced geography's entire disciplinary scope. Given the schools' popularity and impact (in that they had an important influence on how geography was taught in schools), Semple's lectures were a significant platform from which to communicate her philosophy to audiences beyond the academy.

Herbertson's research interests were, to an extent, allied with those of Semple. His 1905 paper 'The major natural regions' was an important manifesto for a systematic approach to geography which, in considering the classification of regional environments based upon climate, vegetation, and topography, might usefully expose the relationship between human society and the geographical milieu.[40] Unlike Semple, however, Herbertson advanced a more nuanced proto-possibilist perspective: rather than proposing a straightforward causal link between environment and society, Herbertson was satisfied to claim only that the influence of the natural region would 'make itself apparent in human affairs.'[41] Despite Herbertson's 'cautious and balanced' position on environmentalism, he shared with Semple a methodological vision of geography which placed systematic research at its core.[42]

For a payment of £30, Semple contributed a number of lectures on 'Environmental Influences' and the 'Geographical Environment of Man' at the Sixth Biennial Vacation Course.[43] Her associate lecturers included Patrick Geddes, who spoke on the geography of cities, and Herbert Fleure, 'an effective lecturer' (in Semple's opinion) who talked on the geography of Wales.[44] Although Fleure had close intellectual links with Geddes – having invited him to contribute to

Aberystwyth's own vacation course several years previously – it is unclear to what extent the three geographers used the opportunity of the Oxford summer school to exchange views on the purpose and direction of the discipline. In some senses, Fleure, Geddes, Herbertson, and Semple represented a broadly similar Neo-Lamarckian approach to geography. They were also united by a common desire always to consider society in relation to environment; to do otherwise was, in Herbertson's view, 'scientific murder.'[45] Where they differed, however, was in their conceptions of region and in the roles they attributed to biological heredity in relation to societal development.[46]

Although Fleure later 'explicitly rejected Semple's ideas', his dismissal of them was not simply a consequence of his opposition to her environmentalist position.[47] Semple considered Fleure to be a 'modest, gentle, curious soul', and recognized him as a source of 'valuable information and suggestions.'[48] Fleure, for his part, appreciated Semple's scholarship, but found a number of her assertions impossible to credit. As a consequence of his close intellectual connection to Vidal de la Blache, Fleure thought Semple's geographical interpretation of history 'not always very judicious', but did not wish to dismiss her work in its entirety.[49] Although it had been Semple's explicit aim to avoid generalizations, Fleure considered Semple's text to be 'sometimes dogmatic rather than scientifically tentative.'[50] This fact, combined with the book's seeming inattention to biological heredity, was the basis of his antipathy towards *Influences* and the reason why it was used as a set text at Aberystwyth only in a very particular way (see Chapter 5). For Fleure, the underlying problem in Semple's work was that she lacked correct 'anthropological experience', and that as a result her beliefs, although genuinely held, lacked credibility.[51]

In her lectures at Oxford, Semple sought to promote a new approach to geographical research and explanation. She saw the summer school as a way to 'help me further to formulate my ideas', since her experience at Chicago had shown the value of classroom discussion in revising and refining her anthropogeographical perspective.[52] As part of the task of communicating these ideas, Semple had arranged with her British publisher, Constable and Company, to have a supply of *Influences* for sale in Oxford and for the book to be included on the course's recommended reading list.[53] Demand for her text proved strong, and upon her arrival in Oxford in August, Semple discovered that 'all of my seminar students (21) and many of my 200 lecture students had provided themselves with the book.'[54] As a consequence of her charismatic lecturing style, rather than the compelling qualities of her book *per se*, Semple made a highly favourable impression on her students and on the members of the public and University community who attended her presentations. A contemporary newspaper report spoke of her 'stimulating personality', 'eloquent delivery', and 'quiet humour.'[55] Semple's 'quiet voice, with which she can do wonderful things' was an important factor in her approbative reception.[56] The acceptance of her anthropogeography depended not only upon its textual representation, but also upon her embodiment of it.

Combining impressive lantern slides – which she had commissioned whilst in Japan – with a convincing oratory, Semple led her audience, as one witness recalled, 'gently face to face with the Truth [of environmental influence].'[57] The extent to which the credibility of her pronouncements depended upon her convincing and enthusiastic mode of presentation was clear: 'Even if one had read her book, it is always far more inspiring to listen to the spoken word than to read the written one, and Miss Semple has a wonderfully stimulating personality.'[58] Semple's effectiveness in communicating her ideas was seen, not only as a peculiar skill but as a model for the dissemination of knowledge. As was noted, 'She can keep her audience keenly alert for a whole lecture without a single note. The Americans seem to make a special study of the art of imparting their information, which would be a great help to many of our learned men.'[59] As a consequence of her enthusiastic evangelism of her methodological and epistemological ideas, Semple succeeded in communicating the basic principles of her anthropogeography to a number of British school teachers of geography, as well as to students at the School of Geography and to parts of Oxford's academic and lay communities. The initial positive response to Semple's work became formalized by the incorporation of her anthropogeography into the department's curriculum and examinations.[60] Although Semple's output became part, in this sense, of the framework of geography at Oxford, reaction to it – and to anthropogeography more generally – was not constant and always approbatory.

Women's Geographical Circle and the Royal Geographical Society

Following her spell in Oxford, Semple returned to library work in London and to writing up the results of her Japanese research for presentation to the Royal Geographical Society. Again, she was based at the Lyceum Club, and her visit coincided with the establishment there of a Geographical Circle. Through its Circle – admission to which depended upon 'participation in original geographical work' – the Club sought to 'promote geographical knowledge.'[61] The Circle was presided over by Bessie Pullen-Burry (1858–1937), an imperialist explorer and anthropologist.[62] In addition to being an important supporter of the suffragist cause, Pullen-Burry was also later a member of 'The Britons', an anti-Semitic and anti-immigrant political group, which had important links to *The Morning Post* (see Chapter 3).[63] Along with the Circle's Vice-Presidents – explorer/traveller Charlotte Cameron, and Violet Roy-Batty, a close friend of the African explorer Mary Kingsley – Pullen-Burry arranged a luncheon in Semple's honour at the Lyceum on 13 November 1912.[64]

The luncheon, which was 'very well attended', was an opportunity for Pullen-Burry to set out the Circle's aims – one of which was to promote practical training for women engaged in geographical research.[65] In support of the Circle's principles, Semple 'illustrated in humorous fashion the fact that the days are gone when the mere possession of a text-book on the subject was considered sufficient equipment for a teacher.'[66] Semple and Pullen-Burry were united in their desire to encourage geographical work in the field. They wished to claim part of the otherwise manly rhetoric of science which had, since at least the eighteenth century, emphasized physical exertion and ocular testimony as central to the 'pursuit of scientific truth.'[67] Something of the foundations for this project were laid with the establishment in 1907 of the Lyceum's Alpine Club (which, in 1909, went on to become an independent organization).[68] The Alpine Club's president was Elizabeth Le Blond (1861–1934), a 'Victorian woman of both spunk and discretion who ascended the Matterhorn in long and abundant skirts.'[69] Le Blond had shown how the trappings of gender could change from being markers of a woman's inability to be in the field, to evidence of what she was able to overcome. The guiding influence of women such as Le Blond and Pullen-

Burry created an environment at the Lyceum which promoted travel and exploration as the bases of physical health and intellectual betterment. This was, in part, a component of the wider suffragist movement of which Pullen-Burry and Semple were enthusiastic proponents. Semple's various contributions to the Woman's Club of Louisville, the Kentucky Federation of Women's Clubs, and the United Daughters of the Confederacy had instilled in her an implicit desire for gender equality, and her geographical work became an extension of that wish. Semple's work in the field served a dual purpose: it satisfied the assumption that a direct sensory engagement with an object of study was necessary to secure correct and reliable knowledge of it; and it demonstrated that a women was able to work successfully and systematically in remote or foreign environments in spite of the perceived limitations of her gender. It was Semple's success in these objectives (and the scholarly texts which emerged from them) that led to her invitation to lecture before the Royal Geographical Society on 4 November 1912.

At the time of Semple's presentation of the 'Influence of geographical conditions upon Japanese agriculture' to the Geographical Club of the Royal Geographical Society, it was relatively uncommon for a woman to enter the Society; more so to address it. For a brief period between 1892 and 1893, the Society had admitted twenty-two women, including the explorer/traveller Isabella Bird to its Fellowship.[70] The issue of the admission of women to the Society – 'The Lady Question' as it became known – had been the subject of prolonged debate among the Fellowship.[71] One of the 'most strenuous opponents' to the admission of women at that time had been the explorer George Nathaniel Curzon (1859–1925).[72] By 1912, however, Curzon's position had changed significantly. As newly-instated president of the Society, Curzon oversaw the purchase of more suitable (and more expensive) premises near Hyde Park. Faced with this expense, Curzon turned 'an eye to new subscriptions' and came increasingly to the view that the election of women to the Fellowship was justified in both meritocratic and financial terms.

The perceived remunerary benefits of extending membership to women were not, though, advertised explicitly. Curzon chose instead to emphasize the scholarly and exploratory achievements of women. As he noted, 'We feel that in the last twenty years women, have, with increasing ability and thoroughness, vindicated their right to be re-garded as serious contributors to geographical science.'[73] In the time which had elapsed since the initial admission of women in 1892–1893, Curzon reported that 'women [including Semple] have read some of the ablest papers before our society' and had 'conducted explorations not inferior in adventurous courage or in scientific re-sults to those achieved by men.'[74] In a sentence which seemed to speak almost directly to Semple's experience, Curzon concluded:

> they [women] have made valuable additions to the literature of travel, and have been invited to lecture in our great Univer-sities; above all, as research students and as teachers, they enjoy opportunities for which they are at least as well equipped as men, and which render them a factor of great and growing importance in the diffusion of geographical knowledge.[75]

Curzon believed that women satisfied not only 'the "emerging stand-ards" of scientific exploration or fieldwork, but also the 'standards' of race, class and gender' associated with membership of the Society.[76] Whilst his position on this matter was resolved during the middle part of 1912, his exposure to Semple and her work helped to cement – or, at least, to render less disingenuous – his thoughts as to the scientific and geographical contribution of women. Given that Curzon had, twenty years previously, been of the opinion that women's 'sex and training' made them 'unfitted for exploration', his revised position represented a radical change.[77] Curzon had encoun-tered Semple's work first in a gentlemen's club, where he 'took up a book which was entitled "The Influence of Geographic Environ-ment".'[78] He recalled his impressions thus: 'This book was written by an author who was evidently a master of the subject with which he or she dealt. It was written with a great knowledge of the subject, no inconsiderable powers of reasoning, and a most agreeable style.'[79] Whether Curzon registered surprise when he discovered the gender of the book's author is uncertain. What is apparent, however, is that he greeted Semple's appearance before the Society with alacrity.

The qualities which Curzon had identified in Semple's text were echoed in her presentation to 'the most fashionable [audience] in London', which took place in the lecture theatre of Burlington House in Piccadilly on 4 November.[80] Dressed in 'a light-blue evening gown with a string of fine pearls' – her attire a marker of difference in otherwise male surroundings – Semple projected calm authority.[81] With only occasional reference to her notes, and with the aid of hand-tinted lantern slides, Semple spoke confidently and engagingly on the relationship between climate and agriculture in Japan – using this example to illustrate her anthropogeography. The *Daily Express* spoke of the 'remarkable spectacle of a woman lecturer holding an audience of some of the greatest living scientists spellbound for more than an hour.'[82]

That Semple was in command of her subject was evident when, in the discussion which followed her talk, Lionel William Lyde (1863–1927), an economic geographer at the University College London, interjected with a criticism of her interpretation of Japanese agriculture. Lyde had a certain reputation for 'making startling and provocative assertions' of this type, and Semple responded in like spirit.[83] Quoting from a statistical account, she retorted 'You have taken a sentence from page three and applied it to page sixteen', at which point Lyde 'rose in almost trembling apology ... and the audience applauded.'[84] Semple's clarity and composure in response to Lyde's bumptiousness impressed Curzon. He concluded the evening's session with an effusive expression, which spoke not only to his opinion of Semple's work but also to the debates then current within the Society about the admission of women to the Fellowship:

> We have had an unusual experience to-night for in the place of the somewhat cautious compliments that are usually addressed to the reader of the paper, we have listened to a series of searching questions put by an intrepid professor [Lyde], and responded to by Miss Semple with a spirit and ability that has given us all the greatest delight. Three things struck me chiefly about the paper: first, the extremely keen and observant eye which Miss Semple must have directed to the objects of her inquiry; secondly, the wonderful beauty of the slides she showed us, many of them from photographs taken by herself; and, thirdly, her unusual power ... of deducing from the phenomena of material existence large generalizations and scientific laws.[85]

A few hours before this successful address, Semple had dined with the Council of the Society at Oddenino's Imperial Restaurant on Regent Street. As one newspaper reported, Semple was only 'the second woman in the world to whom has been shown the honor of being a guest at a council dinner.'[86] Her predecessor had been Isabella Bird. Semple was the only woman among a dinner party of thirty men, and this resulted in an 'amusing incident' when a 'portly butler came to the door and announced to the assembled guests: "Lady, gentlemen, dinner is served".'[87] At dinner, Semple sat at Curzon's right hand next to Major Leonard Darwin (1850–1943), son of Charles Darwin. Darwin was then chairman of the Eugenics Education Society, and was an enthusiastic promoter of social progress by means of the improvement and selection of hereditary traits.[88] Although Semple was inclined to attribute greater influence to environment than to heredity in the development of physical and mental traits, her position on mediated or directed heredity is less certain. Asked later what she talked about with Darwin and Curzon, Semple replied 'I didn't, they did, about themselves.'[89]

Darwin was one of a 'prominent group' of Society Fellows who were then active in advocating the admission of women, and it seems likely that Semple's presentation, if not her conversation at dinner, would have confirmed Darwin in his supportive opinion.[90] Darwin had preceded Curzon as Society president, and his quiet persistence on the matter of female membership – combined with Curzon's own reappraisal of the scholarly and geographical contribution of women – was responsible for Curzon's decision in November 1912, only days after Semple's address, to issue a circular to Society members 'promoting the election of ladies as Fellows.'[91] The fact that this motion came from Curzon, formerly a passionate opponent of female membership, did not go unnoticed in the press. For *The Scotsman*, Curzon's conversion was 'symptomatic' of wider changes at the Society, where 'only a few weeks ago the ... new lecture session was opened with a paper on the economic geography of Japan, by one of the ablest geographers of the day, Miss Ellen Churchill Semple.'[92] With the newspaper press and the majority of the Fellowship in support of the resolution, it was passed successfully, and women permitted to become members from January 1913. Of the 163 women elected that year, at least three – Bessie Pullen-Burry, Charlotte Cameron, and

Violet Roy-Batty – were members of the Lyceum Club's Geographical Circle with whom Semple had dined in November 1912.

By chance, rather than by design, Semple was at the focal point of an important change in the institutional structure of geography in the United Kingdom. Her approach to geography in terms, particularly, of scholarship and work in the field illustrated that gendered assumptions of what it meant to do geographical work were changing. For much of the first half of Semple's professional career women were, to varying degrees, excluded from the discipline's mainstream. In part, this exclusion reflected established notions of what counted as suitable scholarly and scientific pursuits for women.[93] The conduct of science in the field – particularly where it necessitated physical exertion, risk, or, simply, remoteness – was understood as a 'heroic, manly endeavour.'[94] Combined with the exploratory achievements of *various* women travellers, Semple's scientific work in the field was evidence that such undertakings were not exclusively a male preserve.

In some senses, Semple's work had to succeed not only in terms of its scientific value and rigour, but also in its ability to transcend the gendering of knowledge. In achieving the former – in part by being co-opted in the defence of the discipline – Semple's work secured the latter. As one contemporary newspaper recorded, 'It is satisfactory to know that a woman [Semple], by her writings, which occupy the highest rank in recent geographical literature, and by her research work, should be so successful a pioneer in a new and most important branch of geographical science.'[95] The relative enthusiasm with which Semple's anthropogeography was greeted in Oxford and London was, however, a function both of text and of speech: Semple's literary style and impressive locution were fundamental to the successful communication of her work. Whilst the written text was, in most cases, sufficient to satisfy questions about Semple's method, reasoning, and deduction, the fact that she was able to give voice to her anthropogeographical ideas on the floor of the Burlington House lecture theatre, and to defend them successfully in the face of criticism and pejorative opinions as to her gender, lent additional authority to them and to her.

The visual representation of Semple's material similarly mattered to its communication. The relationship between author, reader, and text was, to an extent, replicated in the context of Semple's slide lectures, where a 'performative triangle ... of speaker, audience, and image' facilitated the dissemination of her environmentalist ideas.[96] The particular role of the visual image in this situation was to collapse the geographical distance between the lecture theatre and the field. Semple's photographs 'transported viewers across space' and, for a brief time, made the distinct spaces of field and lecture theatre virtually colocational.[97] As a consequence of the perceived authority of photography as a virtual witness, Semple had the ability to link her anthropogeographical claims to what could be construed of as their visual proofs.[98] Images were one important component of the 'rhetorical triangle' through which Semple's anthropogeographical knowledge moved.[99] Although underpinned by a common triumvirate of producer, object, and receiver, Semple's written texts, her spoken addresses, and her visual representations each communicated something different (or, at least, spoke in different ways to different people) about her anthropogeography.

Although it is possible to dismiss the relative importance of these last two modes – the oratorical and the visual – in the dissemination of Semple's ideas, they had a significance that was disproportionate when compared to the relatively small number of people who were witness to her lectures. In part, this was a consequence of the press reporting which accompanied Semple's visit to Britain. In much the same way that her literary style appealed to certain outlets of the popular press that had reviewed her book, so too did her confident and effective oratorical abilities. Style was certainly not more important than substance, but the manner in which Semple's knowledge was communicated mattered to the ways in which, and by whom, it was received. For parts of geography's professional mainstream, Semple's lecture to the Royal Geographical Society remained a topic of conversation for several weeks; it seemed to speak not only to the question of gender in geographical work but also to the position of anthropogeography and regional description in the discipline. As one member of Semple's audience, the surveyor-geographer Henry George Lyons (1864–1944), later recalled, Semple's method of taking 'the general view of a whole district' was one from which 'there is a great deal to be learned.'[100]

Whilst the Oxford summer school and the Royal Geographical Society were largely professional spaces, Semple addressed a more diverse audience during a popular lecture tour of Scotland under the auspices of the Royal Scottish Geographical Society. At George Chisholm's invitation, Semple travelled north from London to address the Royal Scottish Geographical Society in Edinburgh, Glasgow, Dundee, and Aberdeen. As a guest of Chisholm, Semple was honoured with a 'handsome reception' in Edinburgh to which the University's students of geography were invited.[101] Her tour of the Society's regional branches began in Aberdeen on 19 November, where a 'large audience' gathered to hear 'Miss Semple, who had come all the way from Kentucky to lecture to them – (applause).'[102] What was particularly significant for this audience was that Semple was 'by extraction Scotch, as her name showed, having come from Renfrewshire – (applause).'[103] Semple's address, 'Japan as a type of island environment', used her recent experience of that country to illustrate the principles of her anthropogeography. Japanese agricultural practices and vernacular architecture, among much else, were shown to be the direct consequence of environmental conditions. The following evening, she delivered the same address before a 'very large audience' at the YMCA Hall in Dundee.[104] A local journalist described in detail the format and delivery of Semple's presentation:

> A number of excellent coloured slides were then shown. The audience were treated to some remarkable mountain pictures, showing graphically the hilly nature of the land. The pictures led the audience from the flat seashore to the wooded mountains, showing on the way the cultivation of rice at different stages of growth; the interculture of beans or millet with barley, the case of the mulberry tree and the lumbering industry. In addition to this some splendid pictures of Japanese villages and houses were shown, and an accurate conception of the Jap farmer's existence obtained. Altogether, Miss Semple's lecture was of a most informative character, and we learned many things of the land of the Rising Sun. Indeed, her hour's lecture taught us more than the perusal of many books might have done.[105]

4 The Music Hall, George Street, Edinburgh

Semple concluded her Scottish lecture tour on 28 November at the Music Hall (Figure 4) on Edinburgh's George Street – a meeting attended by an audience of 1,200.[106] The Society's speaker the previous week in Edinburgh had been the Norwegian polar explorer Roald Amundsen (1872–1928), whose topical address – 'How we reached the South Pole' – attracted a similarly large audience.[107] The fact that Semple was able to secure an audience comparable in size to that of Amundsen signals to the particular significance of the public science lecture during this period (rather than any notion of celebrity on Semple's part).[108] Given that the total membership of the Society in 1912 was only 1,898, it is clear that a significant proportion of the audience for her lecture was made up of an interested local public.[109] By addressing the Society, and the public it attracted, Semple was able to communicate her ideas beyond the immediate community of geographical scholars and students to whom they were originally targeted.

The final sentence of the *Dundee Advertiser*'s report – which described Semple's talk as more instructive than any comparable text – points to the peculiar ability of the spoken word, especially when juxtaposed with compelling visual material, to engage an audience's imagination and to convey in a comparatively limited time important components of an argument. The particular site of the provincial lecture theatre – and the fact that the audiences comprised both lay and professional people – served to condition Semple's approach to the communication of her knowledge. She felt it necessary to cast her ideas into an apposite form – to present them as travel narrative, and to make them thrilling and digestible. Whilst her rhetorical style, characterized by extemporaneous delivery, enthusiasm, and subtle humour, seems not to have altered with venue, the content and purpose of her presentations varied depended upon the audience and society to whom she was addressing her work.[110] What Semple chose to say about her anthropogeography was different for student teachers in Oxford; for the elite members of the Royal Geographical Society; and for the interested lay audience of the Royal Scottish Geographical Society. These different conversational spaces demanded subtly different approaches to the communication of her geographical knowledge. More significantly, these venues were also important sites through which the reception of Semple's *Influences* was mediated.

Although listening to Semple talk on the subject of environmental influence was not the same material experience as reading her textual account, both were part of an epistemically-common process through which the transmission of her anthropogeography was enacted. The reception of *Influences*, and the ideas it contained, was a matter not solely of its reading but was also a question of engagements with its other representational forms: the scholarly seminar, the academic discussion, the public lecture, among others. In much the same way that the press had an important mediating influence on the circulation and reception of *Influences*, so too did Semple's own efforts to address her work to both professional and lay communities. Although the difference which her lectures and seminar presentations made to the way in which *Influences* was regarded by its British readers in the second half of 1912 varied, her lectures exerted an important influence on the ways in which the geographical community came to regard both her and her ideas.

SEMPLE'S PUBLIC LECTURES AND SCHOLARLY SEMINARS
IN THE UNITED STATES

By the time of her return to the United States in December 1912, Semple had secured the international scholarly reputation she had for so long cultivated. Her relative celebrity was such that her homecoming was marked by a number of patriotic newspaper reports celebrating her achievements in travel and exploration, and highlighting her positive reception by the British geographical community. Reporting the Royal Geographical Society's recent decision to admit women to its Fellowship, the *Chicago Evening Post* saw Semple as entirely qualified for election:

> We formally propose – let who will second it – the name of Miss Ellen Churchill Semple ... one of three or four students who are developing the comparatively new science which deals with the influence of geographic conditions upon the developments of human society .... We may be wrong, but we know no English woman with superior claims.[111]

For *The Louisville Times*, it was important that Semple's achievements were recognized locally as well as internationally: the paper was keen that Louisville especially should stand as an exception to 'that old Scriptural rule that "a prophet is not without honour save in this country and in his own house".'[112] To the evident approval of the *Times*, 'a sort of intellectual ovation' was arranged by the Woman's Club of Louisville in Semple's honour.[113] Having been seen formerly only as 'an ornament to the more or less frivolous section of Louisville "Society"', Semple had been elevated to the status of 'savant.'[114] Her apotheosis as 'Ratzel's recognized successor' reflected particular credit on her hometown.[115] As the *Louisville Herald* confirmed, the city was 'justly proud to claim for her own a woman of such distinction and learning and charm as Miss Semple.'[116] This local support was not strictly new: ten years earlier, for example, *The Courier-Journal* had described Semple's *American History* as 'truly another feather in the cap of the Commonwealth [of Kentucky].'[117] What had changed in the intervening decade, however, was the fact that knowledge of Semple's work was no longer restricted to 'the more studious' sectors of Kentucky society.[118] As a consequence of its discussion in newspapers and popular periodicals, as well as Semple's teaching and public lectures, her anthropogeography had come to the attention of wider metropolitan and national publics.

Semple devoted much of 1913 and 1914 to the communication of her recent anthropogeographical work in Asia and Europe. The diverse and hectic nature of this programme of dissemination is revealed by the variety of institutions to which she spoke. On 7 March 1913, for example, she addressed the Appalachian Mountain Club at the Massachusetts Institute of Technology in Boston on 'Geographic influences in Japan', before going on to deliver the same lecture four days later before a large audience of 'laymen in geography' at the American Geographical Society in New York City.[119] In both venues, her use of 'Superior stereopticon views' was praised.[120] A little more than a year later, Semple was awarded the American Geographical Society's prestigious Cullum Geographical Medal in recognition of her 'distinguished contributions to the science of anthropogeography.'[121] In accepting this honour, Semple was not only 'the first woman medallist', but was also 'the first person to receive an AGS medal who was not in any way associated with the exploration tradition.'[122] At the award ceremony, attended by an audience that 'filled the large auditorium', the Society's vice-president John Greenough (1846–1934) praised Semple's contributions to both anthropogeography and the disciplinary standing of geography:

> To this branch of science [anthropogeography] the medallist has devoted herself for many years and in many lands with a result truly monumental. Her writings and teachings on the subject are recognized both here and in Europe as authoritative and exhaustive and the charm of style and manner in her books creates a sustained interest such as might not always be expected in scientific material. The catalogue of her works is extensive and our Society honors itself in honoring her.[123]

In response, Semple thanked the Society for a 'rare and signal honour', and for restoring her 'childhood faith in miracles.'[124]

On 26 March 1914, newly-decorated, Semple addressed the Washington, D.C. branch of the Associate Alumnae of Vassar College on her recent travels.[125] The following day, maintaining her hectic pace of dissemination, she lectured to the National Geographic Society on 'Problems of the Japanese farmers'.[126] In each venue, Semple's 'Southern grace and charm' were instrumental in the communication of her anthropogeographical principles – her correct deportment a necessary requirement for the effective dissemination of her ideas.[127] Yet her ability to undertake this peripatetic programme of dissemination depended not only upon her rhetorical abilities, but also upon her relative financial independence. Semple's familial inheritance and royalties from the sale of *American History* and *Influences*, which amounted to several hundred dollars annually, were sufficient to allow her to choose when, and under what circumstances, she undertook paid employment.[128] For a majority of geographers at this time, most particularly female geographers, this was an uncommon luxury.

Although the dissemination of Semple's anthropogeographical work was often facilitated by institutional lectures – at venues such as the Geographic Society of Chicago, the League for Political Education, the Japan Society, and the Geographical Society of Philadelphia – she also made a number of important contributions to the teaching of environmentalism at various colleges and universities. Between 1914 and 1916, for example, in addition to her regular teaching commitment at the University of Chicago, Semple undertook additional lecturing at Wellesley College, Massachusetts; the University of Colorado; and the Western Kentucky State Normal School.[129] Her work in these different institutional settings provides a useful insight into her pedagogical approach, and to the ways in which her research focus had begun to shift to questions of Mediterranean geography.

At the time of Semple's visit in the autumn semester of 1914–1915, Wellesley College was something of a 'female Harvard' – a progressive women's college whose 'stellar cast' of administrators and faculty was exclusively female.[130] The college was one of the few academic institutions in the United States at which female scholars were able to attain academic positions commensurate with their intellectual abilities. At Wellesley, Semple collaborated with Elizabeth Florette Fisher (1873–1941), Professor of Geology and Geography on courses related to environmental influence.[131] Under the auspices of 'Economic and Industrial Geography II' and 'Geography of Europe III', Semple offered lectures on anthropogeography which dealt with 'the influence of the geographic factors of physical environment on man, his industry and his needs; the production of various commodities which supply the needs of man, and the transportation of these commodities.'[132] The courses were structured around 'Lectures and recitations', as well as 'Laboratory and fieldwork equivalent to two hours a week.'[133] In addition to lectures, 'library work' and 'critical discussions' were important pedagogic apparatus in Semple's course.[134] Her enthusiasm and oratorical ease impressed her students, who expressed 'their deepest appreciation of her brilliant work with them.'[135] The students' enthusiasm for Semple's teaching prompted the organization of two additional public lectures: one on 'Japanese Agriculture', the other on 'Militant Germany'.[136] Combined with her 'unusual power of correlation between geography and history', and her 'versatility and exquisite English', the topicality of this second talk provided 'a remarkable insight' into the scope of Semple's anthropogeographical interests.[137]

Fisher – who had invited Semple to contribute to the autumn semester's courses – remained at the college until her retirement in 1926, and some elements of Semple's environmentalism were retained on the curriculum at Wellesley as a result. An appreciation for environmental influence was, for example, also central to the work of Mary Jean Lanier (1872–1961), who joined the college's department of geography in 1917, becoming its head between 1927 and 1939.[138] Lanier had completed her undergraduate and doctoral degrees at the University of Chicago, and through contact there with Semple developed an interest in environmentalist themes. Lanier's environmentalism was expressed most particularly in her 1924 doctoral thesis, 'The earlier development of Boston as a commercial centre', and this perspective directed, or at least informed, her teaching at Wellesley.[139] Lanier had worked closely with Harlan Barrows, and had on a number of occasions taught with him at Chicago's summer school a course entitled 'Influence of Geography on American History, For Teachers of Geography and History'.[140] Despite Lanier's distinctly environmentalist outlook, geography at Wellesley had by the 1940s assumed 'a regional emphasis', and environmentalism was no longer considered an appropriate explanatory approach.[141]

In the summer of 1916, Semple was 'secured for two courses of lectures' by the University of Colorado at Boulder as part of its inaugural summer school in geology and geography.[142] Although a department of geography was not established there until 1927, a number of geographical courses had been offered at Colorado since 1910, and the summer school was an opportunity to place the University's geographical offering on a more substantial base.[143] Semple's six-week courses included 'General Principles of Anthropo-Geography', and 'Geography of the Mediterranean Basin'.[144] Her courses were designed to provide an introduction to the 'various classes of geographic influences and their mode of operation', before grounding them in relation to specific examples.[145] As ever, Semple was keen to demonstrate the validity of her work in the field, and arranged for a 'field study of life under semi-arid conditions' to be undertaken in southwestern Colorado immediately following the summer school.[146]

Environmentalism remained an important component of the geographical offering at Colorado. Virtually every year between 1917 and 1925, the geologist-geographer Walter Edward McCourt (1884–1943), then head of the geography programme at Washington University in St Louis, contributed a course on 'Geographic Influences' to Colorado's geographical summer school. In 1926, responsibility for these courses passed to Ralph Hall Brown (1898–1948), who had been recently appointed to the faculty. Although the course retained the same name, it is likely that its content and purpose was altered subtly. Brown had studied under Ray Whitbeck at the University of Wisconsin, and had inherited from him aspects of the environmentalist tenet.[147] Later, however, Brown's perspective altered as he became convinced by Harlan Barrows' 1922 call that geography should seek to define itself as the scientific study of human ecology.[148] For Brown, Barrows' belief that the objective of geographical inquiry 'should be the study of how man adjusts to the environment … rather than how he is influenced by the environment' became the basis to his later research.[149] It seems unlikely that Brown – in his position as Instructor in Geography at Colorado between 1925 and 1929 – would have employed Semple's text in an instructional capacity. It is certain that he did not when he later taught at the University of Minnesota.[150]

Of Semple's three short-term teaching appointments between 1914 and 1917, her influence seems to have been felt most strongly and persistently at the Western Kentucky State Normal School in

Bowling Green, where she lectured in June 1917. Western Kentucky was a teacher-training institution, and Semple's lecture series was tailored to the specific requirements of school teachers of geography. In addition to general discussion of anthropogeographical principles, Semple offered a number of practical and topical additions. Her lectures included 'Reading the Map of Russia, or France, or Africa, or the Balkan Peninsula, or India' and a discussion of mountain barriers 'with a special view to their effect in the present war.'[151] Two months earlier, Semple had participated in a meeting of the Council of Geography Teachers of Kentucky, which had been organized with the intention of making geography 'more vital and of more abiding interest to the children and teachers of Kentucky.'[152] Semple's presentation attracted a 'large attendance of enthusiastic teachers' as well as Robert Powell Green, who led geographical instruction at Western Kentucky, and who arranged for Semple to present her work there in June.[153] Being a 'Noted author, lecturer and traveller', Semple was a source of particular pride for students and teachers of geography in her home state. Above all her scholarly achievements, the conviction remained that Semple was 'a Kentucky woman.'[154]

Semple's provincial associations with geographical education in the state, and at Western Kentucky particularly, were later honoured on several occasions. In 1929, for example, the Pennyroyal Council of Geography Teachers hosted a dinner in her honour, at which Semple's contribution to geography was celebrated and 'various members of the group told Geography jokes.'[155] Among the Council members then present was Ella Jeffries, who was head of the department of geography at Western Kentucky between 1920 and 1942.[156] Jeffries was later one of the associate members of the Ellen Churchill Semple Geographical Society, which was established by the 'majors and minors in geography' at the Western Kentucky State Teachers College (as the Normal School was then known) in March 1931.[157] The 'Semplia' – the members of the Society – intended in their work to promote geography and to celebrate the contribution of Semple, 'Kentucky's most distinguished geographer.'[158] Alongside the Society's social functions and field trips, a programme of lectures and discussions was arranged to mark 'Semple's career and her contributions to geographic thought and literature.'[159]

Although Semple's scholarly contribution was celebrated at Western Kentucky in this distinctive and enthusiastic manner, her geographical principles were not accepted in their entirety. Jeffries believed that Semple had attributed 'too much to geographic environment', and that Jean Brunhes was correct in his assertion that 'all history can not be explained by geography.'[160] Jeffries did hold to the view, however, that 'it is not possible for us to separate man from his environment', and that the physical environment serves to impose certain restrictions upon societal development.[161] For this reason, she continued to engage with environmentalist themes, and offered a course on 'Geographic Influence in American History' to the 1921 summer school. Jeffries neither rejected Semple's ideas nor accepted them in their entirety. It seems likely that this somewhat considered approach to Semple's work was replicated in Jeffries' geographical instruction at Western Kentucky.

CONCLUSION: PERFORMANCE AND REPRESENTATION

The reception of *Influences* was a matter not simply of how Semple's book was read, but was also a question of the other representational guises which her anthropogeography assumed: the lecture, the seminar, the photograph, the lantern slide.[162] The response to anthropogeography was, in this sense, facilitated both by the reading of *Influences*, and by its communication through a 'network of supratextual discourses' which transcended and transformed the text.[163] Quite what the relative importance of these different communicative modes was in relation to the reception of Semple's ideas cannot straightforwardly be quantified, but it is clear that allocution mattered. What distinguishes the spoken word from the written is the standards by which trustworthiness is claimed and assessed. In her written work, Semple's credibility depended upon her intellectual lineage, her citation of authorities, and her scholarly rigour; in her lectures to the public in Oxford, Edinburgh, and Wellesley, by contrast, it was a question of her palpable enthusiasm, her oratorical skill, and her vivid illustrations.

The dissemination of Semple's anthropogeography hinged upon satisfying the specific measures of credibility, enlightenment, and entertainment associated with the popular lecture. The mobility of her ideas reflected, in this respect, her successful negotiation of local and situated standards of authority. It is evident that the textual, visual, and aural transmission of *Influences'* content demanded of Semple different performative skills, and required of her audience different receptive repertoires. Anthropogeography existed differently (in both a material and an epistemic sense) in its various representational forms and in the different sites of its reception. Although the public communication of anthropogeography mattered to Semple, her specific wish was to address geography's academic community. In the chapter which follows, I consider how *Influences* was engaged with by this audience in its different institutional settings, and how anthropogeography was incorporated into the discipline's curriculum and used pedagogically. Drawing upon evidence of individual reading experiences and teaching cultures at different academic institutions in North America and Britain, I attempt to situate these uses and readings within the context of then-contemporary geographical debates. In so doing, I explain something of the motivating factors which underpinned the teaching of geography at these different institutions, and show how Semple's ideas circulated between them as a function of scholarly networks of personal communication.

5 *Influences'* textbook career

In the years which followed its publication, *Influences* secured an audience, both scholarly and lay, which transcended disciplinary divisions. Semple's book had, though, been designed with one particular readership in mind: university students of geography. Through her lectures at the University of Chicago, Semple had adapted *Influences* 'to students' needs.'[1] She envisioned a clear pedagogical role for the book – the principal function of which was as an aid to education in anthropogeography. In tracing the response of geography's disciplinary community to *Influences*, this chapter considers the different ways in which Semple's book was employed pedagogically, and situates these uses within the context of geography's engagement with environmentalism. From the making of *Influences* at Chicago, through its breaking at Berkeley, to its afterlife at Clark University, this chapter shows that the acceptance and repudiation of anthropogeography depended upon its uneven circulation within a scholarly network defined by its individual members' institutional affiliation, philosophical position, and vision for geography. In tracing a history of this engagement that is partly biographical and partly prosopographical, this chapter's aims are three-fold: 1) to highlight the multiplicity of reading experiences within and between academic institutions; 2) to make clear that the reaction to *Influences* changed through time (and that it did so at different rates, and for different reasons, in different institutional settings); and 3) to show that *Influences* was, in epistemic terms, remade in the different sites of its reading as a consequence of the particular social and intellectual concerns expressed there in regard to the discipline's wider engagement with environmentalism. This chapter begins, as did *Influences*, at Chicago.

ENVIRONMENTALISM AND THE FORMATION OF
THE CHICAGO SCHOOL OF GEOGRAPHY

A concern with environmentalism can be traced at Chicago to the last decade of the nineteenth century, when courses offered there in botany and zoology addressed various aspects of geographical influence. A desire to provide a professional focus to the study of these environmental factors saw the establishment there of a department of geography, proposed first in 1902. The new department's potential scope was set out, in part, by the geologist-geographer John Paul Goode (1862–1932), who proposed three courses exploring environmentalist themes.[2] Goode's approach to human-environment relations was not, as Semple's has been characterized, straightforwardly deterministic: he understood that whilst environmental factors were persistent in their influence, their relative importance in directing the physical and social characteristics of a society would diminish in time as the 'social institutions and conventions' of that society emerged and became dominant.[3] Goode's model of human development was one in which the influence of the physical environment was gradually superseded by that of the social environment. Although Goode had the support of the department's head, Rollin Salisbury, who shared his view that the physical environment was but one factor influencing the development of societies, Goode's proposed classes did not then materialize. Even so, an environmentalist rhetoric – apparent in the 'discourse signalizing' vocabulary of '"geographical influence," "geographic factor," and "geographic condition"' – defined the department's early curriculum.[4]

In the years before the First World War, the principal function of the Chicago department was 'to train men and women for posts in other universities and colleges.'[5] Historical geography formed an important component of this education, and Semple's paper on the Anglo-Saxons of the Kentucky Mountains and her *American History* were required reading – notably for a course on 'Influences of Geography on American History', organized by Harlan Barrows.[6] Barrows had completed his undergraduate education at the Michigan State Normal College in Ypsilanti under Charles McFarlane (1871–1949).[7] McFarlane's perspective on environmental causation – expressed particularly in his later collaboration with Albert Brigham – emphasized environmental causation, whilst also making clear the impact of societies upon their environments. McFarlane's attention to both social and environmental causation was, in some respects, incorporated in Barrows' own perspective and teaching. The intellectual stimulus for his course, which ran from 1904, came principally, however, from two other sources: Semple's anthropogeography and Frederick Jackson Turner's historical geography.[8] Although Barrows subsequently studied under Turner at the University of Wisconsin-Madison for a brief period in 1907, his interest in Turner's work, and also that of Semple, was a consequence of his earlier reading of their works.

Barrows' course on the historical geography of the United States sought to examine 'the geographic conditions which have influenced the course of American history', and in so doing to assess the importance of these factors 'as compared with non-geographic factors.'[9] In attributing rather more significance to human causation than did Semple, Barrows marked out his approach as distinct. What Semple and Barrows did share more obviously, however, was skilful oratory. Like Semple, Barrows 'delivered masterful lectures, beautifully organized. He did not use notes but committed to memory in advance the structure of each lecture and all the figures and illustrations.'[10] Despite Barrows' course initially being offered 'somewhat against Professor Salisbury's advice' (he felt it rather too deterministic), it went on to become 'one of the [most] famous ... of the Department and the University', and attracted large numbers of students.[11] So significant was Barrows' pedagogical reach that, of those students who graduated with a Ph.D. in geography in the United States during the first half of the twentieth century, more than a third were his academic descendents.[12]

Semple joined the department on a part-time basis in 1906 – the first appointment of her professional career, and an indication of the topicality of her geographical interests and her scholarly authority. The combined presence of Semple, Barrows, and Goode meant that students' exposure to environmentalism at Chicago was near universal. Their reaction to it, however, varied considerably. This dispar-ity is illustrated in the differing responses of two of Semple's and Barrows' students – near contemporaries who went on to exert important but distinct influences on the discipline's development: Stephen Sargent Visher (1887–1967) and Carl Ortwin Sauer (1889–1975). Whilst Visher broadly was supportive of Semple's perspective, Sauer was generally (and increasingly with time) critical. In explain-ing why this was so, this chapter shows how – even within a com-mon institutional context – differences in intellectual perspective, classroom exposure to anthropogeography, and application of envir-onmentalist principles in the field, facilitated different readings of *Influences* and thus contributed to the heterogeneous dissemination of the ideas it contained.

Links and lineage: the circulation of anthropogeography

Stephen Visher was raised in a remote agricultural community in South Dakota; his boyhood shaped by 'direct contact with the rigorous regime of the upper mid-latitude continental climate.'[13] Exposure to the 'day to day vicissitudes of the South Dakota natural environment' provided an important background to Visher's later environmentalist concerns.[14] His emergent interest in the role of climate in shaping na-ture and society was cemented in the first decade of the twentieth cen-tury by work with the Chicago geographer-botanist Henry Chandler Cowles (1869–1939).[15] Cowles, a pioneer of plant ecology, conducted research which drew upon environmentalist precepts to define 'a causal relation between plant[s] and environment.'[16] Inspired by Cowles' perspective, absorbed during an ecological expedition to southern Alaska, and by subsequent geological training under Salis-bury, Visher undertook research on the biogeography and regional eco-logy of South Dakota, paying particular attention to the ways in which 'settlers, cowboys, and trappers' had historically adapted to life on the steppe.[17] This research formed the basis to 'The Geography of South Dakota' – a course which Visher offered at the University of South Da-kota between 1911 and 1913. His lectures examined the 'industrial development of South Dakota as dependent upon ... geographic con-ditions, especially location, topography, climate and resources.'[18]

Visher returned to Chicago for doctoral work in geography in 1913, where his ecological training, research in the field, and teaching experience proved useful preparation, particularly for Semple's course in anthropogeography, which he took during the spring quarter of 1914.

Visher read *Influences* for the first time as a requirement of Semple's course. He responded to her lectures with enthusiasm, and considered her text beneficial. He later introduced *Influences* to 'a succession of ... advanced students' at Indiana University, where he taught between 1919 and 1957.[19] Visher's research interests at Indiana focused primarily upon the role of climate, which he deemed 'the most potent' of the 'geographical influences to which man is subjected.'[20] During the early 1920s, he worked closely with Ellsworth Huntington (1876–1947) at Yale University on the research and writing of *Climatic Changes, Their Nature and Causes* (1922).[21] Huntington, a former student of William Morris Davis, was interested in the historical relations between climate and society – particularly in regard to migration and the progress of civilization.[22] Huntington's position on environmental influence was, in this way, rather similar to Semple's. Huntington was eager – perhaps more so than Semple – to advance definitive statements in relation to the role of geographical factors. He called for 'a more precise statement as to the nature and amount, the quantity and quality' of environmental influences.[23] He was critical, however, of Semple's *Influences*, and felt it drew too heavily upon 'book knowledge and not enough from actual observations.'[24] Semple, for her part, thought him 'too obsessed with his climate theory.'[25] Huntington valued the breadth of Semple's scholarship, however, and expressed his admiration in a letter: 'I feel that you must have in the back of your head a complete card catalogue of everything written by several hundred different people.'[26]

At the time Visher took Semple's course in 1914, environmental influence was a predominant geographical concern in the United States and United Kingdom.[27] A survey conducted that year by the economic geographer George Roorbach (who had reviewed Semple's book in positive terms two years previously) found that three quarters of those geographers questioned considered the determination of the influence of geographical environment to be 'one of the chief problems facing the modern geographer.'[28] The geographers who were part of this 'call for research toward the field of geographic influence' were, in some senses, the usual suspects – Ellsworth Huntington, Albert Brigham, and Ray Whitbeck – but their number also included Mark Jefferson (1863–1949) and Lionel Lyde, the economic geographer who had voiced concerns during Semple's lecture to the Royal Geographical Society in 1912.[29] Roorbach's respondents each viewed the prospective contribution that a detailed understanding of geographical influence might make in different ways. For Lyde, for example, it had possible significance for questions of race and 'climatic naturalization.'[30] For Brigham, by contrast, the study of environmental influence was an important basis to 'a more rational definition' of geography as a scientific enterprise.[31] These different interpretations show that, whilst environmentalism was a common concern for geographers in 1914, their conception of it – in terms of its cognitive content and of the work it could do – varied considerably. Semple's anthropogeography was, then, only one of several distinct environmentalist rhetorics, but, as we shall see, whilst the exact purpose of work on geographical influence was differently imagined, *Influences* was read in such a way as to lend credence to these distinct interpretative positions. This is true particularly in relation to the book's pedagogical role in universities and colleges, where it was employed in a number of (often subtly) different ways according to the personal concerns and research interests of faculty members. We can see quite what this meant by considering the use made of *Influences* at one institution in particular – Denison University in Granville, Ohio.

In 1915, the year after Visher first encountered *Influences*, Semple's book was read by George Babcock Cressey (1896–1963), a freshman student at Denison.[32] Cressey's first-year course was organized by Frank Carney (1868–1934), a geologist by training, who incorporated aspects of Semple's thesis into his teaching.[33] Carney had been educated at Cornell University, where, in addition to completing his doctorate, he served as an instructor in the geography summer school between 1901 and 1904.[34] At Cornell, Carney had come under the influence of Ralph Tarr, one of Semple's important supporters.[35] In his work at the geography department's summer school, Carney also met other geographers concerned with environmentalism: most notably Albert Brigham and Ray Whitbeck.[36] Carney's exposure to environmentalist debate at Cornell influenced his later work and, as professor of geography and geology at Denison between 1904 and 1917, he published several articles addressing environmentalist themes.[37] Carney also developed a lecture course on 'Geographic Influences', designed to examine

> several types of geographic influences, as observed in the habitats of primitive peoples, in the development of ethnic groups, in the growth of ideas concerning the size and shape of the earth, and in map-making; in the social, industrial and political activities of advanced peoples, and the influence of topographic and climatic environment on mental and moral qualities.[38]

Carney's course – which was 'innovative and difficult, yet popular' – ran for several years at Denison, and was also offered as part of the summer session of geography at the University of Virginia in 1911 and at the University of Michigan in 1912–1913.[39]

Although Semple's *Influences* was recommended reading for Carney's course, it is unclear quite how many of his students actually read it. As George Cressey later recalled, for example, 'I doubt that we read much of the book, but the ideas were built into ... [the] course.'[40] Carney was an enthusiastic and effective lecturer, and, like Semple, used 'his own extensive collection of lantern slides to illustrate his lectures.'[41] Carney employed Semple's book principally as a supplement to his own teaching. Those of Carney's students who subsequently read *Influences* did so, then, in a way that was conditioned by their exposure to his representation of its content. Carney saw the study of environmental influence as an important component of the 'treatment of human ecology' – that is, the study of different social groups in relation to one another, and in relation to their environment.[42] This ecological perspective in turn influenced Cressey's later research in Asia which focused upon 'the problems of man's use of the land and his habitat.'[43] By the 1930s, if not earlier, Cressey was of the view that the intellectual contribution of *Influences* had been superseded by emergent work in human ecology – a field of inquiry which he regarded as a 'temperate successor to Ratzel and Semple.'[44] Cressey's conception of environmental influence can, in this way, be seen to have depended upon an academic lineage which can be traced from Tarr, through Carney, to a mediated representation of Semple's *Influences*.

Having left Denison, Cressey completed doctoral work in geology under Rollin Salisbury at Chicago in the early 1920s, and there met Semple. Although he did not work with her at that stage, he was somewhat in awe of her reputation and later recalled the 'thrill in lending Miss Semple my fountain pen.'[45] After an extended period of research and teaching in China, Cressey returned to the United States to complete a second doctoral degree (in geography) – first at Harvard, then at Clark University under Walter Elmer Ekblaw (1882–1949). At Clark in 1929, Cressey attended 'what was probably Miss Semple's last class' in anthropogeography.[46] There, fifteen years after his initial encounter with her work, Cressey reread *Influences*. He was struck, more than anything, by Semple's scholarship: 'What impressed me then, and what stands out in the book, was her very extensive documentation. She drew on a vast literature for her references.'[47] Cressey was not, however, uncritical of Semple's approach, and believed that 'she seemed to be more interested in evidence to support her theories, rather than in searching for true relations.'[48] This dubiety as to the value of Semple's work was reflected in Cressey's assessment, on the fiftieth anniversary of its publication, of the influence of *Influences* on the development of human geography. He wrote, 'As to the significance of INFLUENCES, I would certainly say "stimulating". Few of us today would subscribe to her determinism, but she unquestionably opened up many ideas. Judged in terms of the early twentieth century, the values were positive; measured today [1961] the ideas are negative.'[49] Cressey passed the remainder of his career at Syracuse University, where, as this chapter will later show, *Influences* was engaged with in particular ways.

In his thirteen years at Denison University, Frank Carney produced 'seventeen professional geologists and geographers', Cressey included.[50] Although he instilled in many of them an interest in geographical influence, their opinion as to the value of Semple's anthropogeography varied considerably. Cressey's slight suspicion of Semple's position was, for example, in notable contrast to the opinion of his near contemporary, Kirtley Fletcher Mather (1888–1978), who graduated from Denison in 1909.[51] Mather had begun his undergraduate education at Chicago under Salisbury and Atwood, but transferred to Denison in 1907, where he worked closely with Carney and absorbed his environmentalist perspective. Mather's return to

Chicago for graduate work between 1909 and 1915 again brought him under the direction of Salisbury and Atwood and exposed him more directly to Semple's influence. Mather eventually returned to Denison in 1918, and took over Carney's teaching load, including the class on 'Geographic Influences'. The course – which remained 'semi-popular', and had 'a large enrolment' – continued to employ Semple's *Influences* as a required text.[52]

From Denison, Mather transferred in 1924 to the department of geology and geography at Harvard. The following year, he was invited to contribute expert testimony to the trial of John Thomas Scopes (1900–1970), a Tennessee high school teacher who had been arrested for teaching evolution theory (in contravention of a recently-passed prohibitionary law).[53] Despite being a committed Baptist, Mather did not consider the theory of evolution antithetical to his religious beliefs. In written testimony to the court in defence of Scopes, he expressed his view that the paleontological record confirmed 'that life has progressed through time.'[54] Mather's environmentalist education under Cressey at Denison, and his geological work in the field with Salisbury and Atwood, had made clear to him certain of the natural pressures by which biological adaptation was encouraged.[55]

The career of Frank Carney makes clear the importance of personal networks and of institutional hubs in the communication and circulation of environmentalist perspectives in the first quarter of the twentieth century. The relatively peripatetic nature of academic lecturing at that time, and the prevalence of summer schools, means that it is unwise to view any academic institution as homogeneous and unchanging in its engagement with Semple's anthropogeography. For this reason, it is difficult to make general claims about what environmentalism meant, for example, at Yale or Cornell or Chicago or Denison. What is clear, however, is that the foundations for Carney's environmentalist concerns were laid at Cornell – the result of the influence of Ralph Tarr, and of the visiting geographers Albert Brigham (Colgate University) and Ray Whitbeck (State Model School of New Jersey). All of this transpired, of course, before the publication of *Influences*, and it was not until Carney was at Denison that he was able to incorporate Semple's text into his teaching of environmentalism (although it is likely that he previously had employed her scholarly articles).

In addition to the direct influence of Carney's teachings upon Mather and Cressey at Denison, his contribution to summer schools at Michigan and Virginia brought his representation of environmentalism to a geographically-broader audience. In this way, his lecture courses facilitated the transmission of Semple's ideas, but, as Cressey's recollections make clear, the extent to which Carney's teaching corresponded to the exact content of Semple's book was quite varied. In some respects, Carney's promotion of anthropogeography and use of *Influences* was representative of the different ways in which Semple's ideas moved between places – that is, not simply in the material representation of the printed text, but also in various modified and embodied forms. The reception of *Influences* did not, as was made clear in Chapter 4, depend solely upon its reading, but also upon the communication and representation of its content in other forms. That Cressey did not read Semple's book in its entirety in 1915 whilst taking Carney's class does not mean, for example, that he did not receive it. He did, albeit in a modified way.

The significance of Cornell and of Ralph Tarr in the promotion and circulation of environmentalism in the United States in the first two decades of the twentieth century is evident not only in relation to the work of Carney and Whitbeck, but also that of John Lyon Rich (1884–1956). Rich studied geology at Cornell under Tarr and inherited from him the interest in environmental influence that Whitbeck and Carney had previously acquired.[56] In 1911, Rich joined the faculty of the newly-established department of geography at the University of Illinois at Urbana-Champaign where, over the next seven years, he developed courses dealing with, among other things, environmental influence.[57] From 1913, Rich offered a course on 'Influences of Geographic Environment', which dealt with 'The influence of geographic factors ... on his [man's] mode of life, his industries ... modes of communication; the bearing of these factors on historical movements and on the development and policies of nations.'[58] The environmentalist content of this course was mirrored in his 1917–18 course 'Human Geography' – itself concerned with the 'Influence of topography, climate, and other physiographic factors on human life and history.'[59] This course was based upon Rich's attempts quantitatively to analyze the extent to which 'the location and distribution of cultural features, such as towns and clearings in the forests, [are] controlled by topography.'[60] Rich's study identified a strong correlation between topography – meaning relief, slope, and exposure – and the location of settlements and agricultural practices. He saw this as 'the beginning of the quantitative study' of environmental influence, but recognized that it was the work of psychology to 'determine its effect in moulding character.'[61]

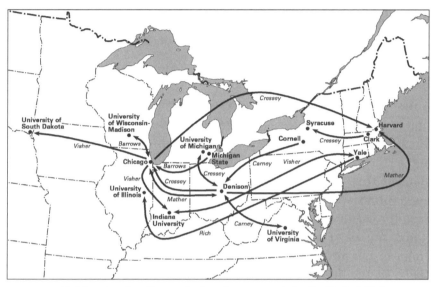

5 Map showing the professional movements in the United States of some of Semple's principal proponents.

When Rich left Illinois in 1918 to pursue a career as a petroleum geologist, the nature of the geography curriculum there changed significantly: 'the program did not survive intact', and Rich's replacement in the post 'taught only 3 or 4 of his formidable array of courses.'[62] In Rich's absence, environmentalism no longer figured to the same extent in the geography curriculum. It is evident that the individual interests of scholars, and their personal research concerns, were important in shaping the ways in which geography was taught at the institutions to which they were affiliated. For this reason, one cannot separate the ways in which environmental themes were engaged with at Denison or Illinois from the personal influence of Carney or Rich (Figure 5). The different staging of environmentalism at these academic hubs reflected the interests and concerns of their students and faculty, rather than some intrinsic quality of those institutions. The geography of environmentalism can be conceived, then, not as a function of the institutions in which it was proposed or opposed, but as an expression of situated and personal engagements which characterized individuals' responses to questions of environmental influence. The influence of personal conviction in response to Semple's work, and to environmentalism more generally, was, as we shall see, apparent particularly at the University of Chicago.

Carl Sauer and 'the rebellion against determinism'

Although the circulation of Semple's anthropogeography depended upon the use of *Influences* in the teaching of environmentalism at different educational institutions, the contribution of her own teaching and lecturing cannot be underestimated – particularly when it can be claimed that 'most of the second generation of American-trained geographers were her students.'[63] Although the important role of Semple's oratory in the convincing presentation of her ideas has been noted, not all of her students were straightforwardly intellectual disciples of her anthropogeographical cause. This was true, most especially, in the department of geography at Chicago, which – despite being an important centre of environmentalist geography and the forum in which much of the content of *Influences* was presented and revised – was also the site where 'the rebellion against environmentalism' began.[64] Chicago produced Semple's most enthusiastic proponents, but also generated her most impassioned opponents. Of this latter group, one student – Carl Sauer – came to reject her work so fulsomely that he was considered thereafter synonymous with 'criticism of environmental determinism.'[65]

Sauer (Figure 6) entered Chicago in 1909. He was a relative newcomer to geography, having trained previously in geology – first at the Central Wesleyan College (Warrenton, Missouri), then at Northwestern University (Evanston, Illinois).[66] His undergraduate training exerted a significant influence upon his later intellectual outlook, but it also had the immediate consequence of rendering him an 'informed dissenter' at Chicago, particularly in respect to anthropogeography.[67] Sauer came by degrees vociferously to oppose environmentalism; he regarded

130

6 Carl Ortwin Sauer, undated

it as an unsophisticated mechanical theory of behaviour. During the 1920s, he offered a 'detailed and devastating refutation of the thesis of environmentalism', which he considered unduly dominant in American geography, and 'singled out for critical consideration' Semple and likeminded contemporaries.[68] Quite why Sauer responded in this way was, as this chapter explains, a question of his scholarly training in geology and ecology, his classroom introduction to anthropogeography at Chicago, and his attempts to apply that theory in the field as 'apprentice and journeyman geographer' variously in Illinois, Missouri, Michigan, and Kentucky.[69]

In 1910, having completed a year of study at Chicago, but before having read *Influences*, Sauer was sent under the titular supervision of Salisbury to complete a geological/geographical examination of the upper valley of the Illinois River.[70] With little direction from Salisbury as to his research focus, Sauer chose to examine the physical origin of the grassland environment and the historical influence of the plains upon pioneer settlers.[71] In this later respect, his study was 'an attempt to apply the orientation then prevailing of human adaptation to physical environment.'[72] Even at this stage, however, Sauer harboured 'some early doubts that this direction [environmental influence] was adequate or proper.'[73] His conviction that the apparently deterministic basis of environmentalist theory was flawed, echoed to some degree Barrows' attempts to present a modified version of environmentalism – one which would 'distinguish between geographic and non-geographic factors.'[74] As has been noted, 'Barrows stood definitely for revision of the environmental doctrine and against extreme determinism. For him, "adjustment" and maladjustment to environment were undetermined except by human choice.'[75] Despite Barrows' modified position, Sauer became increasingly doubtful as to the value of 'the environmentalism tenet', particularly as his experience of work in the field increased.[76]

Whilst conducting fieldwork for his doctoral thesis, 'The geography of the Ozark Highland of Missouri' (1915), Sauer became increasingly aware of the difficulty of making straightforward connections between topography and its physical and social corollaries:

> not all the soils were derived from the weathering of the underlying rock; vegetation paralleled only in part the stratigraphy; kinds of people and their habits did not sort out by physical environment. It was important to know the different terrains, but it was apparent that these only helped to understand the different ways of life.[77]

His fieldwork brought him in contact with various cultural groups – 'German immigrants ... anti-slavery New Englanders ... hill folk from Tennessee and Kentucky' – each of whom 'carried on the usages of their own very distinct and different traditions.'[78] It gradually became clear to Sauer that 'Cultural geography ... was more than "response to natural environment"!'[79] That said, Sauer undertook this and subsequent fieldwork with certain environmentalist presuppositions in place. The extent to which the intellectual influence of Barrows, Semple, and Goode continued to inform Sauer's work at this stage is evident from his field notebooks, which demonstrate an underlying concern for environmentalism. His notebook for 1914 is scattered, for example, with comments which reflect this student training in geographical influence: 'The people are the typical Missouri hillfolk';

'typical backwoods – log cabins – log everything'; 'Fertile population still largely of French descent & decided Frenchtypes. They speak a very broken lingo & are said not to be able to read the printed French. – Good example of influence of isolation'; 'Geog[raphic] infl[uence].'[80]

Sauer came to realize, however, that each observation which appeared to confirm the role of environment in shaping social organization was offset by another which seemed contradictory. The most significant of these observations was that the architectural, agricultural, and social traditions of different immigrant groups in the Ozarks persisted, despite the fact that they were in (culturally speaking) a new environment. If geographical conditions were truly the predominant mechanism for determining these cultural expressions, then it might be assumed that descendents of Germans, French, New Englanders, and Kentuckians would work the land in similar ways, and adapt their architectural practices to reflect the requirements of the environment, rather than maintaining the traditions of their cultural heritage. The fact that the Ozarks were geographically relatively homogeneous, yet culturally were heterogeneous, gave Sauer pause. As previously suggested, however, Sauer's concerns emerged only gradually; it took extended methodological debate and revision before he rejected environmentalism in its entirety. As he later noted, 'Most of the things I was taught ... as a geographer I had either to forget or unlearn at the cost of considerable effort and time.'[81]

Part of the process of unlearning occurred at Chicago in the years between Sauer's field seasons in the Illinois River valley and the Ozarks, when 'a vigorous group of graduate students' began to discuss alternatives to the then-dominant environmentalism.[82] It was in this somewhat critical context that Sauer read *Influences* for the first time, and subjected the ideas it contained to sceptical appraisal. Chicago's nonconformist students – 'young in years and strong in hope' – included Wellington Downing Jones (1886–1957), who began graduate studies there in 1908.[83] Like Sauer, Jones worked primarily under the guidance of Salisbury and was also afforded 'exceptional opportunities for field study' – most notably a two-year spell in Patagonia with the geologist Bailey Willis (1857–1949).[84] In the course of his fieldwork, Jones sought to focus upon 'objective data and inherent qualities rather than on imagined causes [e.g., environmentalism].'[85]

In collaboration with Sauer, he made 'juvenile attempts to select categories for observation' which formed the basis to a methodological paper published in the *Bulletin of the American Geographical Society* in 1915.[86] One of the 'Hints on observation' that Jones and Sauer offered in that paper stated: 'Because of the complexity of conditions in most cases, generalizations must be made with extreme care and only after much accurate observation. The geographer needs to guard against emphasizing geographic influences at the expense of non-geographic ones.'[87] The implications of Sauer's field experience were apparent in the advice which he and Jones provided on examining the characteristics of social groups; they recommend that geographers attend to the

> influence of environment, with special reference to different development of different stocks in the same environment, and to survival of traits and institutions acquired in a previous environment (a fundamental geographic problem of great complexity, the interpretation of which requires great care and in many cases cannot be undertaken).[88]

Although Jones and Sauer were somewhat tentative in their suggestions, their paper was a tangible manifestation of 'the rebellion against environmentalism' then emerging among parts of the graduate community at Chicago.[89] It was several years, however, before Sauer 'rejected definitely the hypothesis of mechanical causation in human affairs.'[90] When he attained a university appointment – at the University of Michigan at Ann Arbor in 1915 – the general introductory course he developed was still 'built about Ellen Semple's ideas as expressed in her *Influences of Geographic Environment*.'[91] Sauer also offered a course on 'Geographic Influences' at Michigan's summer school from 1916 – three years after Frank Carney had addressed the same topic there.

The rebellious mood which Sauer and Jones promoted was not shared by all geography graduate students at Chicago. Bernard H. Schockel, Mary Lanier, Almon Ernest Parkins, and Mary Dopp – whose periods as graduate students coincided with those of Jones and Sauer – were important in promoting environmentalism in their later teaching careers: Schockel at the Indiana State Normal School (Terre

Haute); Lanier at Wellesley College; Parkins at the University of Missouri (Columbia) and George Peabody College for Teachers (Nashville, Tennessee); and Dopp at various high schools in Chicago. Parkins (1879–1940) had previously studied under Mark Jefferson at the Michigan State Normal College in Ypsilanti (where Barrows completed his undergraduate work) and inherited from Jefferson an interest in the influence of human society upon the environment – Jefferson being intellectually opposed to environmentalism.[92] At Chicago, Parkins came under the influence of Barrows and Salisbury, more so than Semple, and this was reflected later in his lecturing at the George Peabody College, particularly in his course 'Influence of Geography on American History, with Special Emphasis on the South'.[93] Parkins remained at the College until his retirement in 1940 and at intervals invited contributions to the College's summer session from former Chicago colleagues. Mary Dopp, for example, offered 'Influence of Geography on American History' on a number of occasions.

The fact that the principles of anthropogeography were tested at Chicago so vigorously was a consequence, in part, of Semple's approach to its teaching. As one student recalled, 'she often said she hoped her work would prove or disprove the values of anthropogeography' and 'that if a better theory came along she hoped she would have strength of mind to embrace it.'[94] In this respect, Semple permitted and facilitated an environment which was productive of critical and independent thought. It was at Chicago, then, that the rebellion against environmentalism in the United States found its earliest expression.

FROM CLASSROOM TO WAR ROOM:
GEOGRAPHY 'AT BREAKNECK SPEED'

When Robert Platt began graduate study in geography at Chicago in 1915, the emergent schism within the department in regard to the question of geographical causation was obvious.[95] Platt – who later came to head the department – had trained in philosophy at Yale and discovered geography by accident. Onboard ship between China and the United States, he fell into conversation with the brother-in-law of Wellington Jones, whose description of geography as then practiced at Chicago interested Platt.[96] What appealed was the notion that 'geography had the advantage of going more to the field for direct observation instead of going to the library to read about things no longer visible.'[97] At Chicago, Platt studied under Barrows, Walter S. Tower (1881–1969), Goode, and Salisbury. Like Sauer, Platt identified most closely with Salisbury's perspective – sharing his view that geographical causation was a valid explanatory mechanism only when considered in conjunction with social influences.[98] Salisbury encouraged his students to 'look into the subject [environmentalism] to see what was in it ... hoping that somebody would investigate a field that might have potential significance.'[99] What Salisbury strove for at Chicago was balance in its approach to environmentalism. Set against the extremes of 'Geography without influences' and 'Geography all influences', Salisbury saw (and hoped the department did too) the physical environment as 'one factor', and thus the value of thinking in terms of 'geographic influences, not controls.'[100] Given the antithetical position occupied by several of the department's faculty and graduate students, Salisbury's wish did not match the reality.

Platt took Semple's course in anthropogeography, for which he was required to read *Influences*, in the spring quarter of 1916.[101] He developed a personal admiration for Semple and thought her 'one of the most stimulating & inspiring' of the university's lecturers.[102] He felt, moreover, that her anthropogeography 'served [a] useful purpose', if taken as indicative rather than definitive.[103] Despite his personal affection for Semple, Platt went on to become 'one of the most eloquent adversaries' of environmentalism – describing it as 'a pseudo-scientific sanction of vulgar belief.'[104] The revision of Platt's perspective on Semple's work, which he saw as both 'stimulating and irritating', was, like Sauer's, a consequence of work in the field.[105] In 1919, a year before gaining his doctorate, Platt was appointed to the department's faculty, and, in addition to regular lecturing commitments, offered an annual summer fieldtrip for graduate students. Whilst the initial purpose of these fieldtrips was to 'gain direct insight into reciprocal associations involving people, space, and the social and physical settings', this objective 'changed somewhat as Platt's own views changed.'[106] He became 'increasingly sensitive to human organization of space' as he found the generalizing principles of environmental theory inadequate to explain the 'micro-conditions and events' he encountered in the field.[107] Platt's intellectual collaborator in this revision was Derwent Stainthorpe Whittlesey (1890–1956) who had joined the faculty in 1919 (the year Barrows assumed the role of chairman from Salisbury), having completed a master's and doctoral degree (awarded 1920) at Chicago.

7 Derwent S. Whittlesey and Robert S. Platt
at the Harvard University graduation of Platt's son, 1954

Although Whittlesey had undertaken his graduate work under the auspices of the department of history, he came to Semple's attention in 1914 when he took her course on anthropogeography.[108] As a consequence of the material Whittlesey encountered in Semple's lectures, he was 'inspired to a lifetime of geography.'[109] Semple and Whittlesey enjoyed a close personal friendship – he called her 'Nennie' or 'Ole Miss' – but he did not share her perspective on questions of environmental influence.[110] In terms of his intellectual concerns (notably the development of quantitative field methods in geography), Whittlesey (Figure 7) was more closely aligned with Jones, Sauer, and Platt, and later came to be 'heavily influenced by the social perspective of the French school.'[111] Although Whittlesey regarded Semple as 'the outstanding representative of the creative imagination' in geography, he, like Sauer and Platt before him, found the environmental ideas she advanced did not correspond with his observations in the field.[112] As this chapter will go on to show, the work of Whittlesey and others came to represent 'the antithesis of environmental determinism.'[113]

Semple's anthropogeography was beginning to be tested at Chicago by the end of the second decade of the twentieth century, even as her professional standing within the United States' geographical community was in the ascendant. In addition to her peripatetic lecturing, Semple had begun to formulate ideas on the historical geography of the Mediterranean region.[114] Her special Mediterranean experience found a practical outlet during 1917 when the United States entered the First World War. Before committing to military action, President Woodrow Wilson emphasized the importance of planning for post-war peace. To that end, in September 1917, Wilson comissioned his advisor, Colonel Edward House (1858–1938), to institute a committee of experts – the Bureau of Inquiry for the Peace Terms Commission, or 'The Inquiry' as it became known – to analyse and advise on the requirements and implications of peace.[115] House drew together a team of academic experts under the command of his brother-in-law Sidney Edward Mezes (1863–1931), president of the City College of New York. Keenly aware of the important role geography might play in any peace negotiations, Mezes invited Isaiah Bowman (1878–1950), director of the American Geographical Society, to join the Inquiry as Chief Territorial Specialist, and to organize a contingent of geographical advisors.[116] Among the geographers who were invited to join the Inquiry in its headquarters at the American Geographical Society were Mark Jefferson, under whom Bowman had studied at the Michigan State Normal College in Ypsilanti, Douglas Wilson Johnson (1878–1944), then lecturer at Columbia University, and Lawrence Martin (1880–1955), who had previously worked closely with William Morris Davis, Tarr, and Whitbeck.[117]

As a consequence of work at Harvard with Davis, and at Yale with Tarr, Bowman was familiar with environmentalist thought.[118] When he joined the faculty at Yale in 1905, for example, environmental influence was already an established component of the curriculum: Herbert Ernest Gregory (1869–1952), a former student of Davis, offered a popular course in the geology department on 'Environmental Influences on Man'.[119] Whilst at Yale, Bowman pursued research in population distribution and regional geography which engaged with the then-dominant environmentalist rhetoric.[120] In 1908, he contributed to the summer session at the University of Chicago and there met Semple. The two thereafter maintained an occasional correspondence, and in 1911 Bowman was among those scholars to whom Henry Holt sent a complimentary copy of *Influences* upon its publication.[121] Given Bowman's familiarity with Semple's work, it is unsurprising that he urged her to 'break any engagement and drop all other work' to contribute to the business of the Inquiry.[122]

Semple joined the Inquiry in New York City in December 1917, and was commissioned to complete a study of the Mediterranean and Mesopotamian regions, setting forth 'principles for partitioning these areas to achieve maximum self-determination, based upon sound consideration of physical and cultural factors.'[123] She worked first as an assistant to Douglas Johnson, before going on to work with Mark Jefferson.[124] Semple's resultant briefings were read by President Wilson at the Versailles conference. Despite the necessarily fast-paced environment of the Inquiry, which required Semple, and its other contributors, to work 'at breakneck speed', she found the work stimulating and appreciated the opportunity to apply her research skills for immediate practical benefit.[125] When, at the close of 1918, the Inquiry members, 'their assistants (together with the materials they had gathered) and numerous officials, including President Woodrow Wilson' departed for the Paris Peace Conference, Semple was obliged to remain in the United States.[126] Despite Semple's extensive previous experience of travel in Europe, she, and other women on the Inquiry staff, were forbidden from travelling, simply 'because they were women.'[127]

Whilst in New York City, Semple was invited to lecture to the 1918 summer session of Columbia University. Environmentalism had been taught there for more than two decades under the 'super determinist' direction of Richard Dodge, an important early supporter of

Semple's work.[128] Dodge had retired from his post at the Teachers College of Columbia University in 1916, but his environmental perspective persisted under the next generation of geographers, at both the College and at the University proper. Dodge's successor at the College was Charles McFarlane, who previously had taught at Ypsilanti, and who maintained Dodge's focus on environmental factors. Environmentalist matters were not, however, restricted to geographical teaching at Columbia. At this time Franz Boas, whose early concerns about the validity of anthropogeography were detailed in Chapter 2, was professor of anthropology. Among the courses he offered was 'Anthropology 1' – an introductory lecture series which dealt, among other topics, with 'the types of man as determined by race and environment.'[129]

Aspects of environmentalism remained a component of Boas' work, but were refined and underpinned by a possibilist philosophy, which saw the environment as important only 'insofar as it limits or favors activities.'[130] Boas 'did not dismiss the environmental factor altogether', but considered it – much like Salisbury, Barrows, and others did – as one influence among many.[131] What Boas was opposed to was strict environmental determinism rather than the consideration of environmental factors *per se*. The year after his death, Boas' library – which included work by, among others, Semple, Ratzel, and Spencer – was purchased by Northwestern University. Boas' copy of *Influences* contains uncut pages – a revealing indication that he did not, in his own edition at least, read Semple's work from cover to cover.[132]

In 1919, the year after Semple contributed to Columbia's summer school, the geographical curriculum again gained an environmentalist imperative under the guidance of the economic geographer Joseph Russell Smith, a recent appointment to the faculty.[133] Born into a Quaker agricultural community, Smith trained initially as an economist before turning to geography.[134] In 1901, he travelled to Europe and, at Leipzig, spent a year 'ostensibly studying' anthropogeography under Ratzel.[135] Like Semple, Smith was enraptured by Ratzel, and described him as 'the finest man in the institution.'[136] Despite his enthusiasm for Ratzel's work, Smith's German was 'scarcely adequate for his needs', and, as a consequence, he 'derived much of his deterministic outlook from the works of Ellen Churchill Semple.'[137] Smith developed his perspective on economic and commercial geography at the Wharton School of the University of Pennsylvania, where he assumed the post of instructor of commerce vacated by John Paul Goode in 1903. Smith's work at Wharton focused principally upon economic geography and industrial management, but retained an environmental component. Smith – in collaboration with Walter Tower (under whom Robert Platt later studied at Chicago) – encouraged an attention in his lectures to 'the mutual interrelationship between earth and man.'[138] Smith's perspective was, then, one of influence rather than determinant.

One of Smith's students during his time at Wharton, and his assistant between 1911 and 1919, was George Roorbach. Both Roorbach and Smith had important connections with *Influences* – Roorbach reviewed the book in positive terms for the *Annals of the American Academy of Political and Social Science*, and Smith, like Bowman, received a complimentary copy of the book from Semple upon its publication.[139] Smith accepted the chair in economic geography at Columbia in 1919, where he continued to organize courses in which the 'environmental factors that influence man's economic and social development' were an important component.[140] The content of these courses reflected and directed the 'man-environment orientation' of geographical work at Columbia during the 1920s.[141] Semple's book played an important role in Smith's teaching: the text was used not only to extend the ideas communicated in his lectures, but also as the basis to various assignments, one of which required students to identify on a map the location of various physical features (rivers, passes, mountain ranges) mentioned in Semple's text.[142]

Smith's enthusiasm for Semple's work, and his promotion of it at Wharton and Columbia, was a consequence, in part, of his positive experience under Ratzel at Leipzig. It had also to do with his interest in economic and commercial geography. Along with Edward Robinson, George Chisholm, and George Roorbach, Smith's principal research concerned economic and commercial geography. It was in this aspect of human social organization that the influence of geographical factors – location, access to resources, barriers to trade, *inter alia* – was most apparent. As Smith made clear in a methodological pronouncement on economic geography, the subject attended fundamentally to 'those geographic influences that affect the economic status of man.'[143] Smith's position was not that the environment necessarily determined the economic development of a society, but that it imposed certain opportunities and limitations which might be differently negotiated. As he framed it, 'Economic geography is the description and interpretation of lands in terms of their usefulness to humanity.'[144] In this way, although Smith, Robinson, Chisholm, and Roorbach were associated with different institutional as well as national traditions of geography, they represented – as a consequence of their particular research interests – a common interpretative community. They engaged with Semple's book in broadly positive terms because it was compatible with their economic concerns.

SEMPLE'S APOTHEOSIS AND THE DECLINE OF ENVIRONMENTALISM

The establishment at Clark University in 1920 of what was termed 'a great geographical institute' was, to some degree, an expression of confidence in disciplinary geography in the United States.[145] In need of a suitable geographer to administer the proposed graduate school, the university authorities approached Wallace Atwood, who had trained as a geologist under Rollin Salisbury, lectured in geography alongside Semple at Chicago, and taken over from William Morris Davis at Harvard upon the latter's retirement. During negotiations over the conditions of his appointment, Atwood was persuaded also to assume the presidency of the university – an enviable position from which to organize the new graduate school.[146] Atwood had the opportunity to choose from the leading geographers in the United States in fulfilling his vision for geography at Clark (although several, including Oliver Edwin Baker (1883–1949), Curtis Fletcher Marbut (1863–1935), and Homer LeRoy Shantz (1876–1958), turned down offers of permanent employment). The first geographer Atwood approached was Semple. Although Semple's work was then subject to criticism by certain of her students and colleagues at Chicago Atwood's was not an illogical choice; having recently been elected president of the Association of American Geographers, Semple was at the peak of her professional standing. Atwood's hiring of her was, in this respect, 'a coup comparable to acquiring a used Rolls Royce in good running condition.'[147]

At Clark, Semple 'came into her full powers as a teacher and director of research.'[148] Clark provided a congenial environment in which Semple could benefit not only from the personal and intellectual support of her colleagues, but also from 'the challenge of training serious graduate students in geography.'[149] Under an agreement with Atwood, Semple taught only during the first semester of each year, and devoted the second to research. Because she was unmarried, and had no children, she was paid $500 less per year than her male colleagues – a remuneratory discrimination she did not discover for several years, and one which prompted her to disinherit the university.[150] The Clark authorities justified this financial inequity on the basis that Semple was 'without dependents.'[151] She considered this a 'mid-Victorian argument from a group of modern capitalists.'[152] It was also a potent reminder to her that, 'though I worked longer hours and made a larger scientific literary output every year than the men professors in my department ... [and had a] national and international reputation [which] equalled or surpassed theirs', the academy, like the society it reflected, was not a meritocracy.[153] Semple's experience proved two things: that 'only exceptional women could find a place on a university faculty', and that the work of a female scholar was not judged or remunerated on its merits alone.[154]

In her decade-long career at Clark, Semple offered courses dealing, in various ways, with environmentalism: 'Influences of Geographic Environment'; a seminar in 'Principles of Anthropogeography' (which she continued also to offer at Chicago until 1923); the 'Geography of the Mediterranean'; the 'Geography of Europe'; and 'Geographic Factors in the Location and Development of Cities'.[155] As at Chicago, Semple taught a number of students who would later achieve professional and disciplinary prominence. These included Esther Sanfreida Anderson (1891–1976), the University of Nebraska's delegate at the International Geographical Congress in Warsaw in 1934, and staff member on the United States Government's War Production Board; Ruth Emily Baugh (1889–1973), first female professor of geography at the University of California; Meredith Frederic Burrill (1902–1997), Executive Secretary of the United States Board on Geographic Names, and later president of the Association of American Geographers; Walter

Ekblaw, editor of *Economic Geography*; Edwin Jay Foscue (1900–1972), lecturer in geography at the Southern Methodist University in Texas; and Preston Everett James (1899–1986), lecturer in geography at the University of Michigan (1923–1945) and at Syracuse University (1945–1970).[156] Semple sought to instil in these students the qualities of thorough research, reasoned argument, and elegant communication for which she herself strove. As one student later noted, 'To think clearly and to express oneself directly and forcibly were her cardinal requirements.'[157]

If Semple's appointment to the faculty at Clark was the high-water mark in her lecturing career, then her selection as first woman president of the Association of American Geographers in 1921 represented her professional apotheosis. Semple had been preceded in the position by a number of geographers who had engaged with environmentalist themes, including Ralph Tarr (1911), Albert Brigham (1914), Richard Dodge (1915), Charles Dryer (1919), and Herbert Gregory (1920). Semple's presidential address to the Association's meeting in Washing, D.C. – 'The influence of geographic conditions upon current Mediterranean stock raising' – combined her emerging Mediterranean concerns with her longstanding anthropogeographical work.[158] Although her presidential contribution was not the last in the Association's history to deal with environmental influence, it did – ten years after the publication of *Influences* – mark the beginning of the end of the dominance of environmentalism in American geography. As is clear from the different historical engagements with questions of environmental influence in the United States, the move away from Semple's anthropogeography and its allied perspectives represented a 'gradual weakening of the hold of physical determinism' rather than a revolutionary transition which the criticisms subsequently advanced by Sauer and others might suggest.[159]

One of the most significant modifications to the environmentalist position was proposed by the Association's president the following year, Harlan Barrows. Barrows' methodological proclamation – 'Geography as human ecology' – was the material expression of the concerns he had raised at Chicago in relation to 'extreme physical causation.'[160] In Barrows' scheme, geography was defined as 'dealing solely with the mutual relations between man and his natural environment.'[161] For Barrows, the adjustment of human societies to their physical environments 'was not caused by the physical environment but was a matter of human choice.'[162] Whilst Barrows' conception of geography was later dismissed as 'a backward step' – since he was seen merely to have replaced an inflexible physical determinism with a rigid cultural determinism – his was among the first in a series of important challenges to the environmentalist position in the United States.[163] The next, and perhaps the most significant, came from Carl Sauer.

Carl Sauer and the 'detailed and devastating refutation of the thesis of environmentalism'

Sauer's time at the University of Michigan between 1915 and 1923 provided him the opportunity to pursue new research concerns and to develop his pedagogic skills. Fieldwork in Michigan, Kentucky, and New England in the early 1920s – mapping '"natural" and "cultural" landscapes' – was important in reinforcing his non-deterministic perspective on human-environment relations.[164] Sauer was commissioned in 1923 by the Kentucky Geological Survey to write a regional monograph on the Pennyroyal Plateau – an area of central and western Kentucky with a characteristic limestone-based karst topography.[165] He was assisted in the project by a Michigan student, John Barger Leighly (1895–1986), with whom he would later work at the University of California at Berkeley. Sauer's investigation of the Pennyroyal, like the Ozarks before it, described a region of near-homogeneous topography, climate, and soil which supported heterogeneous cultural traditions. Sauer travelled and described a region which was populated by the descendents of different settler groups who, despite the particular environmental conditions of the region, 'maintained old ways and attitudes.'[166] This fact served to reinforce his 'growing realization that human activity was the single greatest agent of landscape change and that land use varies according to cultural preferences.'[167]

Sauer's experiences in Kentucky led to the publication in 1924 of 'The survey method in geography and its objectives' – a methodological statement on the scope and purpose of the discipline. In what was, in effect, an expansion of the 1915 paper he had written with

Wellington Jones, Sauer set out in explicit terms his concerns relating to geographical research in environmental influence. For Sauer, it was 'difficult to do scientifically sound work' in the environmentalist mode; studies which operated under the assumptions of environmental influence could only 'throw a half-light on the human scene.'[168] Sauer's new model for research in geography emphasized systematic regional description – a form of chorology – from which relationships between a population and its environment were to be inferred using 'classified and properly correlated observations', but never assumed *a priori*.[169] He sought, in essence, to avoid the 'premature generalizations' he associated with the environmentalist method.[170]

In 1923, Sauer accepted the positions of professor of geography and chairman of the department at the University of California at Berkeley – a move which eventually brought him into contact with the anthropologists Alfred Louis Kroeber (1876–1960) and Robert Henry Lowie (1883–1957), both former students of Boas at Columbia.[171] Lowie in particular encouraged Sauer to reappraise the work of Ratzel – specifically the second volume of his *Anthropogeographie*. As Sauer recalled, 'Lowie got me to understand Ratzel against whom I had been prejudiced by Miss Semple's enthusiasm for her great master environmentalist.'[172] From Ratzel, Sauer derived an epistemic concern which took culture (rather than environment) as an organizing factor. Prior to his refamiliarization with Ratzel's work, Sauer completed a 'sort of habilitation' – a methodological paper which he saw as part of a process of 'emancipating myself from the dictum then ruling at Chicago.'[173] His methodological reappraisal – 'The morphology of landscape' – outlined an approach to geography that placed empirical focus upon cultural landscapes.[174] He explained it thus: 'My object of study is not this fearfully inclusive thing, man, but material culture in areal massiveness.'[175]

Sauer's paper – described later as 'the famous piece that blasted determinism' – sought to disrupt what he saw as the mechanistic and deterministic bases of environmentalism.[176] Although, in this respect, it represented an overt criticism of Semple, it did not affect their 'friendship of long standing.'[177] Semple still considered Sauer 'one of the finest minds that had ever come into my classes', and whilst Sauer admitted to her that his work was in 'quite a different direction from that in which you have worked', he felt sure that she would be 'sympathetic toward what we are trying to do in the study of the succession of natural and cultural landscapes.'[178] Whilst Semple had been subject to similar critical appraisal in France – notably in Lucien Febvre's *La terre et l'évolution humaine* (1922) – Sauer's paper was the most explicit condemnation of her work in the North American literature.

Whilst Sauer's accession to the chairmanship of the Berkeley geography department effectively eliminated the teaching of environmentalism at that institution, it had between 1910 and 1923 previously occupied an important place in the curriculum. The principal exponent of environmental influence at Berkeley was Sauer's predecessor, Ruliff Stephen Holway (1857–1927).[179] Holway had trained as a geologist at Stanford University before joining the geography faculty at Berkeley in 1904.[180] Holway had been converted to geography after attending a summer course at Harvard under William Morris Davis. There, he was exposed to and absorbed Davis' then-strong interest in environmentalism and aspects of Semple's anthropogeography. At Berkeley Holway offered a number of courses which 'mirrored the prevailing geographic opinion of the time' – dealing, in various ways, with human-environment relations.[181] His course on 'General Physical Geography' attended, for example, to 'Land forms, climatology, oceanography, and planetary relations, and their effect upon human affairs', whilst his course on 'Geographical Influences in the Western United States' dealt with 'The geographic conditions which have influenced the exploration and early settlement of the west and the present effect of physical factors on the life of the people.'[182] These themes were expanded, both geographically and epistemically, in his 1918 course on 'Geographic Influences in the Development of the United States', which sought to describe 'the influence of topography and climate of the United States upon location of cities and trade routes and upon man and his activities.'[183]

Although there was an implicit move away from the environmentalist imperative in geographical research under the auspices of the newly-emergent Berkeley school of cultural geography, this did not equate straightforwardly to a rejection of Semple's work. *Influences* was used, for example, by John Leighly for 'many years' in 'Geography 151, American Geographic Thought' – employed as an illus-

147

tration of a particular moment in the historical development of geographical thought in the United States, rather than as a textbook from which to learn.[184] Richard Joel Russell (1895–1971), who joined the faculty on a teaching fellowship in 1920, having previously completed undergraduate study at Berkeley in forestry and geology, read and enjoyed *Influences* 'Many times, 1919–1925, as a student and junior faculty member' – this despite his 'close and continuing friendship with Carl O. Sauer.'[185] Russell's contemporary, Fred Bowerman Kniffen (1900–1993), read Semple's book at about the same time, but responded to it in a different way – a consequence of his longer-standing working relationship with Sauer. [186] Kniffen had completed undergraduate work in geology at the University of Michigan, but had become dissatisfied by geology's lack of attention to human life. In his final year at Michigan, Kniffen came under the influence of Sauer, and accompanied him and Leighly on a summer field trip to Kentucky and Tennessee. There, Kniffen became acquainted with Sauer's emergent dissatisfaction with the environmentalist method.[187] In 1925, he began graduate studies at Berkeley, working closely with Sauer in geography and Kroeber in anthropology.[188] Much like his earlier dissatisfaction with geology, Kniffen felt that 'anthropology neglected the earth', and he became increasingly convinced that Sauer's cultural geography, which drew ever more from the work of Ratzel (at least as it was outlined in the second volume of his *Anthropogeographie*), represented the correct route to explanation in geographical research.[189] Like Sauer, Kniffen described himself as an "anthropogeographer" in the tradition of Ratzel, rather than in the tradition of Semple.[190]

In 1928, Russell joined the faculty of the School of Geology (later the Department of Geography and Anthropology) at Louisiana State University in Baton Rouge.[191] Kniffen, along with a number of Berkeley graduates, followed soon after.[192] Together they formed a department which resembled a 'Little Berkeley', and at which Sauer's cultural geography was emphasized.[193] Although it is unclear whether Russell used Semple's text in his teaching there, it is clear that Kniffen did. *Influences* featured in his graduate seminar 'Elements of Cultural Geography' as late as the 1950s, but was recommended to students only as 'the extreme example of environmental determinism.'[194] Used in this way, Semple's book was presented as part of the historical development of the discipline, rather than as a text from which to learn

148

directly. It was seen to be an outmoded 'period piece' whose 'literary quality was always higher than its scientific quality.'[195]

Although by the late 1920s Semple's *Influences* (and the determinism it was seen to represent) had been effectively dismissed by the geographical community at Berkeley, it continued to be used in other parts of the institution, notably in the Department of Social Institutions. It was there that Clarence James Glacken (1909–1989), an undergraduate, was introduced to the book by Frederick John Teggart (1870–1946).[196] Glacken took Teggart's year-long course 'The Idea of Progress' in 1928, and as preparation for that class read Semple's book.[197] Teggart recommended *Influences* as 'a significant book in the general field of the history of ideas', and Glacken read it alongside 'the *Kleine Schriften* of Ratzel, and the writings of the French possibilist school.'[198] Glacken's encounter with Semple's book – juxtaposed as it was with the work of Ratzel and of the French school – was distinct from those earlier students for whom *Influences* was presented principally as a source of instruction, rather than as a point of comparison. Glacken's approach to the text was, then, somewhat more critical and considered than that of certain of his predecessors. Teggart presented Semple's work 'as an example of environmental explanation of cultural differences ... not as a necessarily valid exposition of the problem.'[199] For this reason, Glacken was inclined to view *Influences* in its intellectual context as 'an important landmark in the history of ideas.'[200]

After almost two decades in non-academic employment, Glacken completed a Ph.D. at the Isaiah Bowman School of Geography at Johns Hopkins University. He returned to Berkeley in 1952, and was appointed to the geography faculty where he inherited Leighly's 'Geography 151, American Geographic Thought'. Throughout the 1950s and 1960s, Glacken continued Leighly's practice of devoting 'at least an hour, often more, to a discussion of selected chapters' of *Influences*.[201] Glacken also made use of Semple's book in his course 'Relations Between Nature and Culture'. As he recalled, 'Students are always interested in some of Miss Semple's more detailed analyses and of course are critical.'[202] This period was marked by the increasing dominance of quantitative methods in geography, and the cultural geography that Sauer had developed in response to environmentalism was itself being challenged. The geography faculty at Berkeley (in an echo of Chicago's earlier schism) was 'divided into factions either defending the "Berkeley School" ... or trying to turn geography's course into a more "modern" direction.'[203] In much the same way that concerns had emerged in the 1920s as to the validity of Semple's method, so too was the authority and value of Sauer's geography questioned in the 1960s.

Concomitant with his teaching at Berkeley, Glacken undertook the research and writing of *Traces on the Rhodian Shore* (1967) – a volume which detailed the development of human conceptions of nature from antiquity to the end of the eighteenth century. An important element of his text was a detailed history of environmentalist thought – tracing its origin from the Classical work of Hippocrates and Aristotle, through Jean Bodin (1530–1596), to Charles-Louis de Secondat, Baron de Montesquieu (1689–1755) during the Enlightenment.[204] The notes which survive from Glacken's unfinished sequel to *Traces* – a volume which sought to address the same themes during the nineteenth and early twentieth centuries – indicate that he saw the environmentalist work of Semple and Ratzel as part of a venerable intellectual tradition.[205] Glacken's approach to *Influences* was, then, one of juxtaposition and contextualization. Having read Semple's book in tandem with its possibilist alternatives, and charted its intellectual ancestry, Glacken understood *Influences* in terms of its contribution to the historical development of environmentalist thought. In his classroom teaching, then, he used *Influences* as illustrative rather than as instructive – it was a text which his students learned *about*, not *from*.

In different ways, and for different reasons, Semple's book had an important role in the teaching of environmentalist thought at Berkeley. Although the principles upon which the book depended had effectively been refuted by Sauer in the 1920s, *Influences* continued to fulfil a particular function. Having gone from being an instructional tool for Holway during the second decade of the twentieth century to an illustrative example of environmentalist thought for Teggart and Glacken from the 1930s, Semple's book fulfilled two distinct roles. In this way it continued to function, albeit in an altered capacity, after its thesis had been gainsaid by Sauer and others at Berkeley. The rejection of *Influences* at Berkeley and at other institutions cannot be regarded straightforwardly as coterminous with the end of its career.

Semple's book had a usefulness that transcended its ability to shape and to direct the course of geographical research.

Semple's students and the promotion and repudiation of anthropogeography

The different uses to which Semple's text was put at Berkeley from the 1910s to the 1960s show the difficulty of identifying a common institutional response to her ideas, or defining one that was uniquely, or notably, Californian. Distinct perspectives on *Influences* within the department at Berkeley were mirrored, moreover, by differences between components of the University of California system. This was particularly true in Los Angeles. The Southern Branch of the University of California, as UCLA was then known, was established in 1919 – an outgrowth of the Los Angeles State Normal School. Geography had been taught at the Normal School since its inception in the 1880s, but it did not attain a dedicated department of geography until 1911. In 1912, Ruth Baugh was appointed to the geography faculty, a position she retained for more than four decades.[206] In 1913 and 1919, Semple visited the Normal School and offered a series of lectures on anthropogeography. She came during this period into contact with Ruth Baugh, and the two established a strong friendship. After completing an undergraduate degree in geography at Berkeley in the early 1920s (prior to the arrival of Sauer), Baugh transferred to Clark University in 1925 where Semple had secured for her a scholarship to complete graduate study.[207] In Semple's view, Baugh was 'probably the strongest woman candidate for a degree that we have had.'[208]

One of Baugh's near-contemporaries at Clark, who had completed his doctoral studies there in 1923, was Preston James.[209] Despite their almost simultaneous encounter with Semple's work, they responded to her teaching in notably different ways. These distinct experiences show that the force of Semple's personality alone did not guarantee the effective communication of her anthropogeography. Before arriving at Clark, James has completed a master's degree in meteorology and climatology under Robert DeCourcy Ward (1867–1931) at Harvard.[210] Ward was an enthusiastic proponent of climatic causation, and the influence of his perspective – along with that of Semple – is apparent in the field diary which James kept whilst visiting Latin America in 1921:

> last night ... we had a fine example of hot house climate. You could wring the moisture out of your clothes and with no wind to evaporate the water, it soon became impossible to indulge in the least exertion .... Even after disease has been eliminated from the tropics, the physiological effect remains to deny for all time any perfect acclimatization of the white races without the loss of the energetic qualities of leadership of that race.[211]

Although James was, as a consequence of his earlier education, inclined towards an environmentalist perspective, he found Semple's teaching of the subject at Clark unsatisfactory:

> Ellen used the book the wrong way. Students had to memorise and repeat what it says, and any attempt to discuss the questions inherent in her philosophy was squashed. I remember how delighted she was one time when I added an example of 'robbers in pass routes.' I told about going through the Mohawk Valley in an automobile as a child, and having the sheriff stop us for going over 8 miles an hour and collecting a fine. This was the only route west in 1911 – and this was truly a robber in a pass route. She loved it. But when I suggested that the Bolivian Plateau might be considered more peripheral than the Atacama (in relation to markets in Europe) she almost threw me out of class.[212]

Despite misgivings as to Semple's pedagogic approach, aspects of her work influenced James' doctoral thesis – 'Geographic factors in the development of transportation in South America' (1923). At some point in the 1920s, however, his faith in the significance of environmental influence appears to have waned. In 1923, he was appointed to the faculty of the University of Michigan (in part on the strength of a letter of recommendation from Semple), where he took the post vacated by the recently-departed Sauer. Although *Influences* appeared on undergraduate reading lists there, the text was not incorporated directly into the curriculum.[213] As James later noted, 'This was an intellectually exciting period, because we were coming out from Semple and joining Sauer.'[214]

The principal spurs to James' reassessment of Semple's anthropo-geography were periods of chorographic fieldwork in Trinidad and New England.[215] These regional studies, similar to those undertaken by Robert Platt, showed how 'several cultures have left their own peculiar impressions in the landscape' – the reverse, in short, of Semple's model.[216] Like Sauer and Platt, it was work in the field – particularly that which revealed cultural heterogeneity in geographically similar environments – which caused James to question the value of an environmentalist approach. In 1929, the same year as James' research, Derwent Whittlesey completed a similar study in New England and coined the term 'sequent occupance' to describe 'studies of the processes of change in the occupance of an area.'[217] As was later noted, 'Studies in sequent occupance [like those of James, Whittlesey, Platt, and Sauer] represent the antithesis of environmental determinism.'[218] James' focus upon regional description and the influence of society upon the environment was not a straightforward rejection of Semple's perspective, but a more considered basis from which to advance conclusions about the relationship between society and the environment. James did not believe that Semple's approach was wrongheaded, just overstated: 'When Ellen wrote about how nature whispered to man the answers of how to get along in an environment, she was letting the poetry of nature get the upper hand.'[219] James' intention, in examining the interrelation of landscape and human life, was to redress the balance somewhat. Commenting later on the influence of Semple's book, James noted 'There are many parts that are just as valid today as Ellen thought they were .... But because parts were so extremely deterministic, the whole book has been set aside.'[220] That said, there was for James enough of value in Semple's book for it to remain on the reading list at Michigan throughout his tenure, and later at Syracuse University despite the 'violence of the anti-environmentalism prevalent there.'[221]

In the three years which separated James' and Baugh's encounters with *Influences* at Clark, Semple's classroom manner had mellowed, and the prescriptive experience which James recalled was replaced by a more dialogic approach. *Influences* henceforth 'was used ... as a basis for the discussion which Miss Semple directed. Students recited,

asked questions, drew out the author on subjects not generally understood, or on points where there were differences of opinion.'[222] After
completing her Ph.D. at Clark in 1929, Baugh returned to the University of California at Los Angeles, where she resumed her teaching
career. There she made considerable use of Semple's book 'in undergraduate courses in which historical subjects and material in human
geography were being considered', and it was placed on reading lists
in both the history and geography departments.[223] Semple's ideas
were also discussed in a graduate course entitled 'Development of
Geographic Thought', taught from 1925 by Clifford Maynard Zierer
(1898–1976), who had studied under Stephen Visher at Indiana
University, and who had completed an environmentalist doctoral
thesis at the University of Chicago (in part under the guidance of
Barrows). Under Zierer's direction the 'departmental philosophy' at
UCLA 'emanated largely from Ellen C. Semple, Ellsworth Huntington,
and others of similar bent.'[224] As a result of the frequent reference
made to *Influences* in Baugh's and Zierer's courses, 'the copies placed
on reserve at U.C.L.A. were well worn.'[225]

The environmentalist orientation of the geography curriculum at
UCLA was challenged, however, on a number of occasions, most notably by Joseph Earle Spencer (1907–1984), an undergraduate between
1925 and 1929.[226] For Spencer, the 'simplistic and one-sided views'
embodied by the environmentalist position 'caused me considerable
difficulty, and I was summarily ejected from class on several occasions for arguing with instructors.'[227] Spencer was supported in his
dubiety by Jonathan Garst (1893–1973), an Iowa-born, Edinburgh-
educated geographer, who joined the faculty in 1927. As Spencer recalled, Garst's views 'were very different from those of the American-
trained faculty', and Garst introduced Spencer to the work of European geographers, particularly those of the French school.[228] Garst
gave a focus to Spencer's concerns as to the value of the environmentalist position and provided him with an alternative methodological approach to geographical research: 'Garst set Joe's orientation
in a nondeterminist direction', and it was at Garst's suggestion that
Spencer later undertook graduate work at Berkeley under Sauer.[229]
When Spencer returned to UCLA in 1940 as an instructor in geography – he brought with him the cultural geography of Sauer, and
this came gradually to replace the environmentalist physical geog-

raphy which previously had dominated at the Los Angeles department.

Baugh remained at UCLA until 1956, achieving full professorship in 1953. Although not uncritical of Semple's work, most of the courses Baugh offered at UCLA were 'on regions and topics that had been of interest to Miss Semple – Europe, Historical Geography of the Mediterranean Region, the Geographic Basis of Human Society.'[230] Whether directly by reference to *Influences,* or indirectly through her own teaching, Baugh facilitated the communication and dissemination of Semple's geographical work during much of the first half of the twentieth century. In addition to her promotion of *Influences* at UCLA, Baugh's most significant contribution to Semple's intellectual legacy was the assistance she afforded Semple in completing her final work, *The Geography of the Mediterranean Region* (1931).[231] Semple's ill health (she had suffered a heart attack in 1929) meant that much of the editorial responsibility for the book was assumed by Baugh.

In a career at UCLA and its predecessor institution which spanned five decades, Baugh, like Semple before her, influenced the undergraduate experience of a number of future geographers – Robert Cooper West (1913–2001), Evelyn Lord Pruitt (1918–2000), and Peter Hugh Nash (b. 1921) among them.[232] Although, as Nash recalled, 'Baugh almost worshipped Semple, and much of this admiration rubbed off on me', Baugh's personal affection for Semple did not translate to an evangelical espousal of anthropogeography.[233] In some respects, Baugh's use of Semple's text contrasted markedly with her Clark contemporary Ekblaw, who 'transferred his admiration and respect for her to her book.'[234] Whilst Baugh had inherited something of Semple's research interests and passion for geography, Ekblaw might more properly be understood as her intellectual primogeniture – her most enthusiastic recipient and proselytizer.

In 1913 Ekblaw had been selected as botanist and geologist for the MacMillan Crocker Land Expedition to the Arctic. Beset by problems, the Expedition was forced to remain in the high Arctic for four years, rather than the one year originally intended.[235] In this time Ekblaw began an investigation of the Inuit population of northern Greenland, a subject which became the basis of his doctoral dissertation – 'The Polar Eskimo' (1926) – which was supervised by Semple.[236] Ekblaw's interpretation of the Inuit culture was 'something close to the stereotypical deterministic viewpoint', and offered 'a nightmare version of environmental determinism' – a notable contrast to Franz Boas' investigation of the Baffin Island Inuit.[237] Ekblaw's adherence to the environmentalist perspective was seen as an 'example of over enthusiasm.'[238] Ekblaw was rigid in his reference to Semple's work, which he used as the sole basis for his 'Human and Cultural Geography' course at Clark, offered until 1948–49.[239] As one student later recalled, 'Criticism of anything was resented', and when an attempt was made to discuss possibilist theories in geography, 'Ekblaw lost his temper and said my remarks were absurd because "You can't grow bananas at the Pole".'[240] Ekblaw's admiration of Semple's book – which he regarded as 'the final word on the subject [anthropogeography] in the English language' – went beyond a straightforward adherence to her ideas; the rigidity of his outlook serving in some ways to misrepresent her ideas, or, at least, to exaggerate them by association.[241] As a consequence of Ekblaw's persistent adherence to environmentalism, a 'latter-day form of environmental determinism predominated' at Clark throughout the 1930s and 1940s – well beyond the endurance of this perspective at other institutions, includeing UCLA.[242] As one student of that period recalled, 'we were expected to be familiar with Ellen Churchill Semple', but 'Vidal de la Blache we viewed only in passing.'[243]

Between 1922 and 1946, almost one third of geography doctorates completed at Clark, among them George Cressey's, were directed by Ekblaw, and the influence of his adherence to environmentalist principles was often apparent in their content: for example, 'The influence of location on the evolution of Duluth, Minnesota' (1933), by George Henry Primmer (1889–1946); 'Geographic backgrounds of Babylonian culture' (1934), by Sidney Everette Ekblaw (1903–1990); and 'Geographic factors in American tung culture (Southeastern United States)' (1943), by Ruben L. Parson (1907–1983).[244] An unintended consequence of Ekblaw's teaching was that a number of his students inherited an exaggeratedly deterministic opinion of Semple's anthropogeography.[245] For one student, George Tatham (1907–1987), it was not until he reread *Influences* in preparation for a contribution to Griffith Taylor's 1951 volume *Geography in the Twentieth Century* that he appreciated the extent to which his view of it had been shaped by Ekblaw's perspective.[246] Here, again, geographical location and social context mattered to the reading of Semple's book: under Ekblaw's tutelage, students received the 'strong impression that Semple was extremely deterministic and Ratzel equally, if not more, so.'[247]

In addition to the cohort of graduate students Ekblaw supervised between the 1920s and 1940s, a larger number of undergraduate students attended his lecture courses and summer seminars. Several – including Albert Sigfrid Carlson (1907–1975) and Stephen Barr Jones (1903–1984) – went on to occupy important positions within the geographical academy in the United States. Carlson, who enrolled in Ekblaw's course in 1928, and who later headed the geography department at Dartmouth College, recalled not only the circumstances in which he read *Influences*, but also the practicalities of his reading: 'I … underlined it, took sentence outline notes on it and, at that time, was able to locate most of the place names in the chapters and explain their geographic significance.'[248] His later evaluation of Semple's book was positive: 'I believe the book as important as Mackinder, Brunhes, La Blache and Huntington's works.'[249] Unlike a number of his contemporaries at different institutions, however, Carlson maintained an enthusiasm for *Influences* and continued to recommend it to undergraduate geographers at Dartmouth as late as the 1960s. It is worth noting, however, that there was something of a tradition of environmentalism at Dartmouth – in 1850, for example, Ira Young (1801–1858), professor of natural philosophy, offered a geographical course based upon Arnold Guyot's *Earth and Man*.[250]

Carlson's compatriot, Stephen Jones, who attended the Clark summer school in 1928, was similarly enthused by his initial encounter with *Influences* – he found it 'extremely interesting and stimulating' and read it 'kiver to kiver [sic].'[251] Jones' enthusiasm was, however, short lived. When he returned to graduate studies at Harvard in autumn 1928, he was forced to defend his newly-acquired environmental perspective to the recently-appointed Derwent Whittlesey. Jones failed and, as he recalled, 'Whittlesey de-environmentalized me.'[252] After completing a period of teaching at the University of Hawaii, Jones went on to found the committee (later department) of geography at Yale University.[253] It was at Yale that Jones became aware of the ways in which *Influences* was regarded by non-geographers. He recalled having Semple's book 'pushed at me (figuratively) … by several social scientists, mostly rather elderly and retired' who disagreed passionately with her thesis.[254] In this respect, Jones saw Semple's book as having exerted a damaging influence on professional geography, since it was seen by those in other branches of the humanities and social sciences to be 'the geographers' bible' – embodying the discipline, its scope, and methods.[255] Having been an adherent of Whittlesey's brand of cultural geography for more than two decades, this interpretation (or misinterpretation) of geography's discipline irked Jones.[256]

As the examples of Baugh and Ekblaw show, the persistence of Semple's book as a pedagogic tool at particular institutions in the United States was, in part, a function of Semple's diaspora in both its first and second generations – that is, the students whom Semple taught, and those her students went on in turn to teach. There was not, then, a simple and uncomplicated transmission of Semple's anthropogeography. As Edward Said has suggested, any idea in the process of its relocation is 'to some extent transformed by its new uses, [by] its new position in a new time and place.'[257] This was true of Semple's anthropogeography. In the different representational guises which her book assumed in the classrooms of Baugh and Ekblaw, it was transformed epistemically and its meaning and implications were mediated. As has previously been suggested, however, whether Semple's book was used in the 'right' or 'wrong' way is not of principal concern. The question of the geography of the reception of *Influences* is, rather, a matter of why the book meant particular and different things to particular and different people, and of how these understandings changed across space and through time. *Influences* did not contain a fixed and canonical meaning which was either accurately or erroneously interpreted by its different readers – its meaning was always, in some senses, in flux as it was remade and negotiated. The career of Semple's book was not determined simply by the acceptance or repudiation of its principal thesis (however defined) but by its very malleability.

In 1939, at the annual meeting of the Association of American Geographers, John Leighly in conversation with Ekblaw 'learned ... with astonishment' that *Influences* was still used as a textbook at Clark.[258] His incredulity did not mean that Leighly considered Semple's book moribund or inconsequential, for, although *Influences* functioned in notably different ways for Leighly at Berkeley and Ekblaw at

Clark (and, indeed, also for their students), it *did* function. Even in its repudiated form, the book served a particular role for Leighly – its career redefined, but enduring. This seeming contrast between the book's material permanence and its epistemic malleability has been summarized thus:

> The 'career' of a book begins, ordinarily, immediately upon its birth, and the most vigorous and vital years are the years of infancy, as was true of *Influences*. A book responds to its environment by multiplying in number of copies more or less proportionately to its ability to make friends and interest people. Qualitatively, however, it remains the same (unless, of course, there are sudden mutations when new editions are published) with a constancy that may be embarrassing to the author and refreshing or disappointing to the reader .... Although untouched for years, as long as a copy exists anywhere a book, like a bear in winter, continues to 'live' dormantly.[259]

The transition of *Influences* from its 'vigorous and vital' debut, to a period of doubt and repudiation, occurred at different rates in different places, and at different rates for different people.[260] As this chapter will go on to show in the context of *Influences'* career in the United Kingdom, a common context for the reading and reception did not mean that the book's trajectory was spatially uniform. Despite the shared disciplinary concerns which informed the initial reading of Semple's book, the uses to which it was put, and how it was viewed with time, were multiple. The career trajectory of *Influences* in the United Kingdom – like that in the United States – although conditioned by the initial context of its reception, was manifest differently in different places and for different people.

INFLUENCES IN THE UNITED KINGDOM

Herbert John Fleure was appointed in 1918 to the newly-created Gregynog Chair of Geography and Anthropology at the University College of Wales at Aberystwyth, where he outlined a syllabus influenced by the intellectual triumvirate to which he subscribed: anthropology, geography, and history.[261] Rooted in what would later be described as a possibilist perspective, his syllabus echoed the French school of regional geography, and was distinct from the environmentalist focus that characterized contemporary geographical instruction in the United States. Fleure's methodological desire, described in his *Human Geography in Western Europe* (1918), was to promote a geography which attended to regions as defined by 'areas on which different men have set their characteristic stamp' rather than by their topography, climate, or ecology.[262] In Fleure's view, it was 'impossible to treat man [simply] as a creature of circumstance.'[263]

Despite Fleure's intellectual orientation – and the doubts he had expressed in his review in *The Geographical Teacher* as to the validity of Semple's method – he frequently recommended *Influences* to his students. One of these, Emyr Estyn Evans (1905–1989), read it in 1923. As he later recalled, Semple's book was presented as a tool by which 'to exercise our critical faculties', rather than a text from which to learn directly.[264] Evans had originally entered Aberystwyth to study Latin, but soon 'fell under the spell' of the inspiring Fleure.[265] His pre-university education had not been explicitly geographical, and, as Evans later noted, 'the title [of Semple's book] appealed to a beginner.'[266] The context for Evans' reading was shaped by Fleure, and he engaged with the work critically, having been warned of its potential limitations. Evans came increasingly under Fleure's influence during his undergraduate study and – with Fleure, Emrys George Bowen (1900–1983), and Harold John Edward Peake (1867–1946) – formed what was later termed 'the "Aberystwyth School" of historically-oriented human geography.'[267] Much of the focus of the school was on 'the racial characteristics, both physical and social, of various peoples, and on their powers of adjustment to particular climatic circumstances.'[268] Whilst the human ecological orientation of this work was superficially similar to that proposed by Harlan Barrows at Chicago, and the cultural geography of Sauer at Berkeley, it was distinct in that it sought to identify the characteristics of particular racial types as the basis to understanding their cultural expression (settlement, agriculture, trade, and so on) in space. For Barrows and Sauer (and so, too, for Semple), race mattered less.

Although the work of the school 'aroused considerable interest', it had 'little impact on regional geography ... in Britain', at least when compared to the influence of Sauer in the United States.[269] More significantly, the geography of Fleure and his disciples was not intended as a corrective to Semple's anthropogeography (as was Sauer's), and, as a consequence, existed in parallel with it for much of the 1910s and 1920s.[270] Evans, who went on to teach at Queen's University, Belfast between 1928 and 1970, employed *Influences* there in much the same way that Fleure had done at Aberystwyth:

> I ... recommend students to read the work, warning them, as I was warned, of its weaknesses. At a certain stage I think it is immensely stimulating and I have not known students to suffer in the long run. It is much more dangerous for students who are not geographers.[271]

Unlike at Aberystwyth, aspects of environmentalism were favourably incorporated into the teaching of geography at the University of Oxford during the tenures of Halford Mackinder and Andrew Herbertson.[272] When the Honours degree syllabus was put before University authorities for approval in the early 1920s, environmental influence featured prominently in both its physical and human components. The proposed course in 'Principles of Physical Geography' promised 'a study of the influence of geographical conditions ... upon man', whilst the 'Geography of Man' considered the 'influence of geographical environment upon physical type and culture.'[273] Around this time, Semple's book – then one of the recommended texts included in the unofficial book list issued to students – was read by John Norman Leonard Baker (1893–1971) in preparation for the Diploma in geography.[274] Baker, who had recently completed undergraduate work in modern history, was 'quite naturally ... critical of it', but appreciated the fact that *Influences* conveyed to Anglophone readers, albeit in a mediated form, 'something of what Ratzel had been writing about.'[275] The value of *Influences* to Baker came from its representation of Ratzel's anthropogeography, rather than from its communication of Semple's own perspective. Despite doubts as to the book's validity, Baker selected a copy as a reward when he was granted the Herbertson Memorial Prize in 1921.[276] Whilst the titular focus of geography at Oxford had not changed radically in the period between Semple's lectures at its summer school in 1912 and Baker's encounter with her text a decade later, the Neo-Lamarckian perspective associated with both Mackinder and Herbertson was no longer in vogue.[277]

Baker joined the faculty at Oxford in 1923, where he spent the remainder of his career. So influential was Baker on the subsequent development of the department that 'for many years, and particularly ... during the 1930s, "Baker" and "Oxford geography" were almost synonymous terms.'[278] During this period, the 'criticism of [Lucien] Febvre and the rise of the "possibilists" had an adverse effect on Semple's book', but, as Baker was keen to point out, this was 'all the more reason for reading it to see exactly what she said!'[279] *Influences* remained, for precisely that reason, 'one of the "recommended" books in our ... list given to undergraduates' until at least the 1960s.[280] Pejorative marginalia in surviving copies of Semple's book at Oxford indicate that those students who chose to read *Influences* during the 1950s and later "to see exactly what she said" encountered it in very different terms than those who had read the book at the time of its publication. Although, as Baker pointed out, 'It would be a mistake to judge the value of the book by present-day standards', student readers during the second half of the twentieth century considered certain of Semple's arguments to be barely credible.[281] That *Influences* was subject not simply to doubt but to ridicule is evident in a marginal sketch (Figure 8) depicting Cloud Cuckoo Land' – the implication here being that the ideas contained within the book reflect derangement or naivety on Semple's part. Although it was Semple's somewhat florid prose that attracted most negative comment – 'What an imagination'; 'This is laughable'; 'Come off it!' – the perceived racism of her text was subject to particular censure.[282] Semple's claim, for example, that despite the vicissitudes of the tropics, the British colonist is able 'to do a white man's stint of work' is qualified by the acerbic suggestion 'i.e. kicking nigs, supping gin.'[283]

A similar transition from initial eager acceptance to later enthusiastic repudiation was apparent too at the University of Cambridge, where environmentalism had formed a core component of the geographical curriculum since the first decade of the twentieth century. From 1903, Alfred Cort Haddon (1855–1940) – 'the great anthropologist of Cambridge University' according to Semple's assessment of him – offered a twice-weekly lecture course in anthropogeography.[284] Haddon, a zoologist and anthropologist by training, had undertaken a number of ethnographic expeditions during the 1880s and 1890s which had convinced him of the important connection between culture and physical environment.[285] His syllabus for the anthropogeography course covered 'The geographical distribution of races according to continents. The influence of geographical environment on the life, arts, social organisation, and migrations of the more important peoples.'[286] From 1907, his course – occasionally taught by his assistants, Miss L. Whitehouse and Alison Hingston Quiggin – formed one of six subjects in the examination for the diploma in geography. When the Tripos, or Honours, programme in geography was prepared in 1919, anthropogeography was, again, one of half-dozen examinable components.[287]

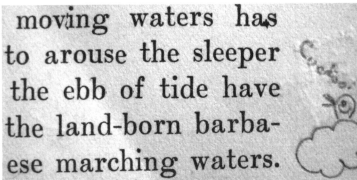

8 Marginal sketch depicting Cloud Cuckoo Land in a University of Oxford copy of *Influences of Geographic Environment*

Semple's book was included on the undergraduate reading list as preparatory material for Haddon's course, and was the first book which Henry Clifford Darby (1909–1992) purchased upon arrival in Cambridge as a sixteen-year-old undergraduate in 1925.[288] He later, and ironically, recalled the experience: 'Without realising that my geographical soul might be imperilled ... I bought ... E. C. Semple's *Influences of Geographic Environment* (1911), and read it with interest.'[289] In his second year at Cambridge, Darby purchased an English-language translation of Lucien Febvre's *A Geographical Introduction to History* (1925) – 'the antidote to the "Influences".'[290] Febvre's book was a monograph on the inadequacies of anthropogeography and environmental influence, and 'The major butt of this attack was borne by the luckless Ellen Semple.'[291] In his text, Febvre set out a 'very vigorous statement' of the principles of possibilism.[292] Febvre understood the physical environment to impose constraints upon societies which, rather than determining a particular course of cultural development, afforded a series of possibilities which could be differently exploited.[293] Possibilism did not, however, eliminate deterministic thinking – it merely changed the causal focus from environment to society.

Following the publication of Febvre's 'outstanding work', *Influences* was, as Darby recalled, 'subject to some pretty severe criticism by my fellow students and myself, to say nothing of our teachers.'[294] At Cambridge, as at Oxford, Semple's *Influences* was no longer encountered favourably. Febvre's critique of the anthropogeographical and environmental perspective, and the emergence of possibilist alternatives, had undermined its credibility – it had become 'outdated and out-moded.'[295] That said, Darby did not consider Semple's text to be without value: 'One might ... say that the "Influences" is a shocking book that misled people and put them on the wrong train. Yet it did provoke us to think and, after all, one enjoyed even its absurdities.' For Darby, 'The book ... filled a gap, and filled a need.'[296] In contrast to Darby's somewhat critical reading of *Influences*, Semple's book enjoyed a rather more positive reception at the University of Liverpool. There, racial geography and anthropology formed part of the course in human geography offered by Percy Maude Roxby (1880–1947), who had studied under Herbertson at Oxford.[297] George Tatham, a second-year undergraduate in 1925, read Semple's book in preparation for this course and found it stimulating. He felt, however, that it did not equal her *American History*, which he had read the previous year.[298] Institutional context was important, then, to Tatham's reading of *Influences*. As he noted later, 'Whenever I have encountered people whose opinion of Geography seems to have been adversely affected by "Influences" it has usually turned out to be a result not so much of the book itself but of the way they were introduced.'[299]

From Herbertson, Roxby had inherited a 'cautious and balanced' approach to environmentalism.[300] Although he was 'deeply aware of the influence of the physical environment' he was equally conscious of the 'long human moulding of the landscape.'[301] In this respect, although Roxby 'appeared as a "possibilist", following Lucien Febvre and Vidal de la Blache', he did not reject the value of Semple's work outright.[302] For this reason, Roxby recommended *Influences* 'as background reading for his Human Geography course.'[303] Tatham, as was noted earlier, subsequently read Semple's book under Ekblaw at Clark University, and had his perspective of the text altered as a result.[304] Since Roxby and Ekblaw understood *Influences* in different ways, and sought to draw different conclusions from it for the benefit of their students, Semple's text functioned differently at Liverpool and at Clark. The fact that Tatham's impression of the book was different in these different contexts (and changed again when he reread the book in the 1950s), makes clear the significance both of time and of location (particularly of institutional circumstance) to the acts of reading and interpretation.

In considering the reception of *Influences*, it is clear that its readers often changed their mind – and changed it repeatedly in Tatham's case – as to its value. It is apparent that the interpretative repertoire with which readers engaged Semple's book was rarely fixed and invariable, but was fluid and mutable. Exposure to new social and intellectual experiences, to new texts and to new contexts, often meant that readers' expectations and interpretative approaches were altered – as *they* changed, so did their reading of Semple's book. The motivations for reading *Influences* varied too. For some, it was read to gain Semple's perspective; for others, it was an entrée to Ratzel. For still others, Semple's book was read to gain insight into the historical development of the discipline, or simply to fulfil a course requirement. These different motivations facilitated different readings.

Despite Febvre's efforts to 'settle the score with geographic determinism', Semple's book remained an important text in undergraduate education in the United Kingdom throughout the 1930s.[305] That Semple's book continued to receive attention was due, in no small part, to the criticism to which it had been subject. Its notoriety served as motivation for Oskar Hermann Khristian Spate (1911–2000), who read *Influences* in 1931 whilst an undergraduate student at the University of Cambridge.[306] The book appealed to him 'because it was ... a standard, full-dress discussion of a problem – environmentalism – which ... has always bulked large in geographical thinking.'[307] Although the popularity of *Influences* had, in large measure, given way at Cambridge to a focus on possibilism and the work of Febvre, Semple's book continued to be read 'by serious students.'[308] Spate saw *Influences* (and, by implication, the work of Ratzel), as having made an important challenge to the 'empiricist and anti- (or at best a-) theoretical tradition in social science.'[309] Although Spate – who later advanced work in probabilism, an intermediate position between determinism and possibilism – recognized the important contribution that possibilism had made in countering 'a lot of pseudo-scientific junk' associated with environmental determinism, he also considered it to have 'put nothing very positive in its place.'[310] Given this, Spate thought that Semple had been 'too totally cast out' by the geographical community, and that there was much in Semple's oeuvre that was worthy of consideration.[311] As he noted, 'there is a great deal in Semple's book to think over, to verify, to discuss, to dispute .... It is like Marxism: invaluable stiffening of one's philosophical bony structure.'[312]

Whether read as an exercise in philosophical ossification, or as a representation of what human geography once was or could have become, *Influences* was, by the mid 1930s in the United Kingdom, understood no longer to be part of geography's contemporary canon. That it continued to be read is apparent, however, from copies of the book which survive in numerous British university libraries. Tangible manifestations of private reading practice (date stamps, marginal annotations, worn pages, and rebinding) describe an irregular but sustained engagement with Semple's book – a material record of in-tangible intellectual interaction.[313] Of these, marginal annotations are particularly revealing. At once a personal commentary and an open proclamation, they attest to the interplay between reader, text, and author. The motivations for these exchanges are not always clear, but their somewhat illicit nature allows their commentary to be more critical than might otherwise be possible. They are, of course, occa-sionally a form of vandalism – a site where frustrations, either with the text, or more generally, can be expressed. In one Cambridge copy, for example, the word Oxford has been altered to read Poxford.[314]

The accusation 'RACISM!' is proclaimed on several pages of a copy of *Influences* held by the University of Birmingham.[315] Whilst this undated comment cannot be taken as a proxy for the initial reception of *Influences*, it is apparent that Semple's ideas had been subject to debate in Birmingham's department of geography since the early 1920s, where Febvre's possibilism exerted a significant influence. Michael John Wise (b. 1918), a student during the second half of the 1930s, was, for example, issued a 'firm warning' in relation to Semple's apparent determinism by the department's head, Robert Henry Kinvig (1893–1969), who chanced upon him reading *Influences*.[316] Kinvig had trained initially as a historian, but was invited by Roxby to join him as an assistant lecturer in geography at Liverpool in 1919.[317] The pair shared a similar outlook and believed that 'Human geography was much more than the study of the influence of the natural environment upon human groups.'[318] It was as a consequence of Kinvig's recommendation to his students 'that Febvre's *A Geographical Introduction to History* (1925) became a much-read book' at Birmingham.[319] In this context, it is unsurprising that Semple's book was subject there to criticism: one anonymous reader described it as 'foolish rot.'[320]

Whilst Semple's expressed concern had been to eliminate race as an explanatory category, it is clear that for several readers encountering the book several decades after its publication *Influences* conveyed quite the opposite message. Annotations in a number of library copies show that, in both the United Kingdom and the United States, Semple's text was considered racist and representative, in an embryonic form, of Nazi geopolitics. An ironic 'Heil Hitler!' appears, for example, in one copy of the book at the University of California at Berkeley alongside Semple's prediction that 'It is impossible to resist the conclusion that the vigorous, reorganized German Empire will one day try to incorporate the Germanic areas found in Austria, Switzerland and Holland.'[321] Swastikas are, moreover, present in the margins of copies held at the University of Chicago and Queen's University Belfast.[322] For one reader at the University of Sheffield, consulting a copy of *Influences* printed in 1947, Semple's discussion was 'racialist', whilst for another at the University of Oxford, her text was methodologically unsound: it 'doesn't embrace falsification principle', it 'relies more [and] more on examples', and thus was 'unscientific.'[323] In another Oxford copy, Semple's suggestion that 'The method of anthropo-geography is essentially analytical' is accompanied by the pejorative suggestion 'I wouldn't have guessed from reading this book.'[324] Of these comments, the last two are particularly telling, since, as we have seen, the positive reception of *Influences* in the years immediately following its publication depended precisely upon its perceived scientific qualities.

That Semple's thesis was seen to afford a nomothetic approach to geographical research had ensured its initial approbation. Yet, its reception was neither spatially nor temporally uniform – its *raison d'être* was undermined by an increased attention to the Vidalian tradition of regional geography, introduced through the work of Febvre and Brunhes. Whilst the British geographical engagement with Semple's notions of environmental influence might be dismissed as transitory, it was more significant. Although Semple's ideas did not dominate in Britain to the extent that they did in the United States, they provided an important framework around which disciplinary geography was constructed. Rather than serve merely as a methodological guide, Semple's book had, particularly for early readers, a totemic significance – its epistemic proclamations were an important indicator of geography's disciplinary remit. Attention to the ways in which *Influences* was reviewed upon its publication in Britain, and to how it was subsequently used, shows its reception to be in large measure a question of its perceived usefulness in outlining a scientific methodology for geography. For Chisholm, and for the *Scottish Geographical Magazine*, *Influences* spoke to a particular moment when the discipline was concerned not only with its academic institutionalization, but also with its epistemic and methodological bases. The reception of Semple's ideas did not depend, however, simply upon a pragmatic assessment of their applicability – affirmation of her perspective was a function of the Neo-Lamarckian approach evident in the work of Mackinder and Geddes, among others. The fact that Semple's ideas could be seen to build upon Spencerian Social Darwinism by advancing a more nuanced multi-causalism meant that her work represented an important contribution towards the Neo-Lamarckian scheme.

Perhaps the most significant challenge to *Influences'* intellectual position in Britain came with the publication of Lucien Febvre's *A Geographical Introduction to History*. At base, Febvre's criticisms related to the generalizing principles with which Semple's ideas were seen to be underpinned. For Febvre, the 'older technique of generalization and comparison' was logically flawed – he advocated, instead, the study of specific geographical regions, and their particular qualities.[325] Whilst the tension between Semple and Febvre seemed to reflect that between a nomothetic and idiographic conception of geography, this was not so. The regional geography which, to some extent, came to replace environmentalism was promoted using the very arguments that had ensured the earlier enthusiastic reception of Semple's book – namely that it would provide for geography 'a method of research ... appropriate and peculiar to it' and one sufficiently rigorous that 'geography may find its logical position among the sciences.'[326] Put simply, the acceptance of Febvre's anti-environmentalism depended not only upon the rejection of Semple's principles, but also the desire to legitimize geography as a science which characterized disciplinary self assessment in the first decades of the twentieth century.

The repositioning of British geography during the 1920s and 1930s from its initial adherence to environmentalist ideas, to a closer alignment with possibilist alternatives (principally associated with the French school) was not an immediate paradigmatic shift, but was gradual, spatially and temporally uneven, and motivated by a number of different factors. Semple's work was not replaced in the affections of British geographers by that of Febvre because it was seen to be self evidently better. It had to do, rather, with the types of questions geographers wished to ask, and with the ways in which they conceived of the relationship between nature and society. As the examples of Fleure and of Kinvig make clear, personality and educational experience mattered to the ways geographers engaged with the work of Semple and Fleure. The reception and subsequent rejection of *Influences* and the ideas it contained was a process that was inherently subjective: it depended upon the judgement and opinion of individual geographers and their students.

Despite the important role of subjective assessment in the relative dominance of environmentalism and possibilism in Britain during the 1920s and 1930s, the book's institutional setting mattered too. At the University of Edinburgh, for example, the use of Semple's book post-dated the retirement of its most enthusiastic proponent, George Chisholm. The fact that *Influences* was subsequently used at Edinburgh by Alan Grant Ogilvie (1887–1954) and James Wreford Watson

(1915–1990) in undergraduate and postgraduate teaching did not mean, however, that they understood it in the same way as Chisholm, or used it for the same purpose.[327] As has been previously described, *Influences* fulfilled a pedagogic role even after its content had, in large part, been repudiated. It is for this reason, then, that *Influences* remained on lists of recommended reading in the United Kingdom decades after its methodological influence and disciplinary topicality had faded.

CONCLUSION: THE RISE AND FALL OF 'THE GREATEST WOMAN GEOGRAPHER'

In the autumn of 1929, with *The Geography of the Mediterranean Region* two-thirds complete, Semple suffered a severe heart attack, complicated by cardiac asthma, and was incapacitated for several months.[328] Her teaching career at Clark was effectively ended. By the following summer – her head 'clear and vigorous' – she resumed a limited programme of work.[329] That autumn, she relocated to a boarding house overlooking the campus of Clark University, and, with the assistance of Ruth Baugh, began to draw *Mediterranean Region* to its conclusion. In the winter, the Association of American Geographers held its annual meeting in Worcester – home to Clark University. Semple, having summoned sufficient strength, attended the meeting and presented her final paper.[330]

In the summer of 1931, on medical advice, Semple left Worcester; moving first to Petersham, Massachusetts (where she completed her Mediterranean book); then to Asheville, North Carolina; before finally settling in West Palm Beach, Florida. Shortly after her arrival, Semple received news that she had been awarded, on the recommendation of her former colleague John Paul Goode, the Helen Culver Gold Medal by the Geographic Society of Chicago in recognition of her 'distinguished leadership and eminent achievement in the advancing of the science of geography.'[331] The Society's minute book described her as 'the greatest woman geographer', and noted that the medal was conferred 'by unanimous vote.'[332] Since Semple was unable to collect the award in person, the Society arranged for her erstwhile Chicago colleague, Charles Carlyle Colby (1884–1965), to present it to her during his 1932 spring vacation in Florida.[333] Recognition of another kind came in the reviews which followed the publication of *Mediterranean Region*. In addition to uniformly complimentary periodical reviews, Semple also received personal congratulations from a number of European geographers, including Albert Demangeon (1872–1940) and Emmanuel de Martonne (1873–1955), to whom she had sent autographed copies of her book.

Shortly before her death, Semple wrote to the President of the University of Kentucky, from where she had received an honorary doctorate in 1922, describing her rapidly deteriorating health:

> I am nearing the Great Divide, whence the final journey will be swift and short. But I was able to play the game to the end – even after the grave figure of Death had established its ultimate claim to me – and to complete my big book on the Mediterranean.[334]

She also used the opportunity to dispatch books and personal artefacts to the Memorial Library at the University, keen that these should be placed in her home state rather than at one or other of the universities at which she had taught. She died on 8 May 1932.

In the four decades which preceded her death, the nature and position of Anglo-American disciplinary geography had changed significantly. Semple's career as a geographer paralleled the discipline's institutionalization and professionalization in the United States, and her contribution to its methodological focus was significant and, for a time, even central.[335] Semple was not, of course, the only proponent of environmentalism during this period, but her *Influences* proved unusually important in communicating the principles of anthropogeography. Given that her book 'was published when few English [speaking] geographers read German', it served also as an important means of presenting to students and scholars of geography a selected part of Ratzel's work.[336] Semple's presentation of anthropogeography coincided with, and helped to define, a period of methodological realignment in disciplinary geography in the United States and United Kingdom. Her book succeeded in fulfilling a pedagogical role associated with this realignment – that of providing 'a firm interpretation of the influence of the environment', written 'in such a way that the [student] reader understands Semple's meaning.'[337] *Influences'* intended function was to provide a complete and coherent statement of Semple's perspective on environmental influence – not a definitive statement on the remit and methods of anthropogeography, but an indication of potential scope and possible approaches. It was, then, in its pedagogical role that Semple's text had its most direct influence upon the teaching of geography during the second decade of the twentieth century, and upon the discipline's subsequent research focus.

In different places, and for different people, *Influences* meant different things, and was put to different uses – evidence of what Livingstone has elsewhere termed a 'reputational geography.'[338] The reading, reception, repudiation, and reappraisal of *Influences* varied with time and across space, and varied between people in (sometimes different) institutional, cultural, and national contexts. As with the reviewing of Semple's book, its use was, in part, 'a collective and institutional phenomenon.'[339] That *Influences* was differently staged in different institutional contexts was a consequence in no small measure of the individuals who comprised the departments of geography, geology, sociology, anthropology, and history at which Semple's book was brought to bear on the environmentalist question. The institutional uses of *Influences*, and their engagement with environmentalism more generally, cannot be separated from the individuals of whom they were comprised. As the examples of Sauer, Ekblaw, Tarr, and Carney make clear, the interests and passions of a leading faculty member – particularly at a time when individual departments of geography (where they existed at all) were relatively small – could dominate the ways in which geography was conceived of, and the uses to which Semple's book was put.

The individual readings of *Influences*, together with the different and particular uses to which it was put in educational settings, confirm that the reception of Semple's text was not a fixed and singular event, but was an ongoing process – changing either in terms of location or time as a consequence of shifting attitudes, novel experiences in the field, or the vagaries of scholarly topicality. Understanding reception as something which, in some ways, both preceded and proceeded the moment in which the encounter between reader and text (or audience member and lecturer) took place has implications for what it means to think about the dissemination of knowledge. The next chapter considers the broader implications of this varied attention to Semple's text by exploring the ways in which it can help us conceive of the nature of the reception of texts and of knowledge, and illuminate the ways in which geography matters to the processes of knowledge creation and dissemination as well as to its criticism and rejection.

6 Reflections on the geography of reception

In the seventy-five years since her death, a process of historiographical conflation has rendered Ellen Semple synonymous with Friedrich Ratzel, and has made her anthropogeography conceptually indistinguishable from environmental determinism. This elision has abridged the rich intellectual history of anthropogeography, reducing it to an imprecise synopsis: 'the Ratzel-Semple environmental-determinism method.'[1] Cast as the 'American disciple' of Ratzel, Semple's position within the disciplinary record has become that of 'the prophet of geographical determinism' or, less charitably, 'the bogey-lady of a slightly silly concept that has now happily been abandoned and forgotten.'[2] The tendency to associate Semple simply with 'environmentalist dogma' and 'what is now known as environmental determinism' – and to view her work only in relation to that of Ratzel – elides the contemporary significance of her geographical contribution, and abridges the venerable tradition of environmentalist scholarship in North America.[3] As this book has shown, Semple's anthropogeography was neither straightforwardly a restatement of Ratzel's principles, nor was it introduced to a scholarly community ignorant of questions of environmental influence. The response of geographers to Semple's book was, in fact, part of a longstanding debate within the discipline (and others) as to the suitability of environmental caus-ation as an explanatory approach.

That Semple features at all in the disciplinary record – when many of her female contemporaries are routinely 'written out of histories of geography' – is, of course, noteworthy.[4] Recent historiographical work has attempted to recuperate the role of women geographers, and has described the important contributions they made in, among other spheres, school and college teaching, exploration, research, and editorial work.[5] Given that Semple's ideas were intimately connected with an important phase of geography's methodological development, she occupies a rare and enduring place in the discipline's canon. She tends to figure only, however, as a representative of (what is seen to be) an outmoded and brief disciplinary flirtation with environmentalism. Underlying this particular portrayal of Semple is a desire to narrate progress in geography's disciplinary history. This progressivist perspective holds that the discipline – as it has matured – has left behind erroneous adolescent enthusiasms, becoming increasingly sophisticated.[6] The notion of progress thus delineated is not one in which development comes from building upon earlier conceptual foundations, but rather in tearing them down and building anew.

Revisionist histories of geography have offered contextually-nuanced accounts of the discipline's development, but relatively little has been done to problematize the synonymous categorization of anthropogeography and environmental determinism. As a consequence, the pejorative associations of Semple's work – not least the perception that it was racist and foreshadowed Nazi geopolitical ideology – persist.[7] Recent concern regarding the emergence of a neo-environmentalist agenda in geographical research has reinforced Semple's role as arch determinist.[8] Semple thus occupies a singular and contradictory position: celebrated for her pioneering methodological contribution, yet simultaneously chastised for her deterministic rhetoric. Whilst it has not been this book's intention to recuperate Semple's reputation (either personally or professionally), it is evident that the disparaging characterization of her intellectual contribution has served to obscure the contemporary significance of anthropogeography. By situating Semple and her work in their disciplinary context, it becomes possible to understand why *Influences* had the impact it did, and why its reception varied in geographically- and temporally-particular ways.

SCALES, NETWORKS, AND THE RECEPTION OF KNOWLEDGE

Semple's professional career, and the propagation of her anthropo-geography, coincided at the turn of the twentieth century with the institutionalization of geography in the United States and United Kingdom. At that time environmentalist principles – particularly as expressed in Semple's formulation – were co-opted to provide a methodological basis upon which the discipline defined its *raison d'être*. Semple was not, of course, the only proponent of environmentalism during this period, but her *Influences* proved unusually important in communicating certain of its precepts. The commendatory initial reception of Semple's written work was contingent upon a number of factors: her scholarship, her intellectual genealogy, her investigative rigour, her literary flourish, and her demonstration of anthropogeography as a field science. These criteria mattered to varying extents to Semple's different audiences, but what underpinned the acceptance of anthropogeography as a suitable focus for geographical inquiry was its apparently 'sound scientific' quality.[9] Although Semple's intention was not to promote anthropogeography as a nomothetic method, its logic and deductive reasoning were, for a number of geographers, 'a good illustration of the meaning and value of scientific geography.'[10]

Whilst a number of readers and critics considered Semple's anthropogeography to be the quintessence of the environmentalist method, the reception of it was heterogeneous: it varied within nations and between them; within institutions and between them. The particular disciplinary schism in British geography in 1912 meant, for example, that the context in which Semple's book was read there differed importantly from that in the United States. In Britain, *Influences'* reception was largely a question of its perceived usefulness in outlining for geography a scientific methodology; it was seen to speak to a particular moment when the discipline was concerned not only with its academic institutionalization but also with its epistemic and methodological foundation. Whilst the potential contribution of Semple's book in this respect mattered also in the United States, her work was viewed there by a number of readers in relation, more generally, to the project of American scholarship. *Influences* was thus rendered there as something akin to a national triumph.

176

The limitations associated with employing the national as a unit of assessment in reception study – such as comparing the American and British readings of *Influences* – are well understood, but the question of 'what the correct scale of analysis is at which to conduct any particular enquiry into the historical geography of science – site, region, nation, globe', remains unresolved.[11] Geographers and historians of science, aware of the problems associated with privileging the explanatory potential of one spatial scale over another, have been reluctant to advance definitive answers to this question. In this book I have sought to show that social networks were centrally important to the circulation, dissemination, and reception of *Influences*, and that they offer a valuable alternative framework for interpreting and explaining the ways scientific knowledge moves and is differently understood in the sites of its reception.

Whilst certain scales – the national, for example – facilitated particular types of engagement with *Influences* (as the example of the Close furore in Britain makes clear), the social networks within which Semple's ideas circulated were typically not defined by a particular scale, but by their trans-scalar qualities: professional relationships, shared perspectives on geography's research agenda, and enduring personal friendships. Whilst certain of these networks can be seen to correspond to a particular geographical scale (the metropolitan in the case of the writers and readers of newspapers reviews, for example), a number (members of a disciplinary community, for instance) cannot be rendered so neatly as cartographical abstractions. For this reason, it is this book's contention that the reception of *Influences* – and of scientific knowledge more generally – is not a question of a correct scale of analysis, but rather has to do with making connections between and across all scales.

In the same way that the social networks through which Semple's ideas moved were not defined solely by geographical scale, the acts of *Influences'* reading were a function of multiple interpretative influences and the unique combination of analytical positions. The fact that *Influences* was read differently by different people, at different times and in different places, unsettles the notion that the printed text straightforwardly permits the faithful reproduction and circulation of the ideas it is intended to represent. The apparent physical immutability of the book has, indeed, been central to understanding how knowledge made in one place moves to another, in its physical guise and it its conceptual form. As has shown, however, the individual copies of *Influences* were neither physically nor epistemically immutable. Rather than an impediment to the communication of anthropogeography, the fact that Semple's book would 'often mutate, creating an enhanced or different understanding' of her message was a necessary and vital part of the communicative process.[12] The reception of Semple's ideas – and of scientific knowledge more generally – is not a function so much of its reproduction, but of its reconstitution. The movement of knowledge and ideas, whether understood spatially or epistemically, results in (and depends upon) modification of that knowledge.

Influences was central to the dissemination and circulation of anthropogeography, but Semple's ideas also existed in a number of different representational guises. The reception of her environmentalist philosophy was not a matter simply of the reading of her book. Anthropogeography was propagated and debated in other media and venues: the scholarly and popular lecture; the newspaper and periodical review; the classroom and field site. Each of these communicative nodes facilitated the mediation of Semple's anthropogeography in different ways. Semple's personal performance of her work – typified by impassioned oratory and captivating lantern slides – communicated something qualitatively different about anthropogeography, for example, than did the classroom discussions of those teachers for whom *Influences* was a pedagogic guide to their engagement with environmentalism. In these different discursive venues, different understandings of anthropogeography were mobilized. These encounters, although superficially distinct from the reading of *Influences*, were part of a common interpretative process. The reception of Semple's anthropogeography cannot be illumined fully, then, by reference solely to the reading of her text. *Influences*, it is clear, mattered to the communication and reception of anthropogeography, but it was not the only thing that mattered.

Although Leah Price's provocative assessment that 'the geography of the book is still making up its rules' was an important prompt to this book, it has not been my intention to codify precisely the rules which might define geography's engagement with print culture.[13] It is apparent, however, that geography's perspective on the circulation of knowledge through print has some important implications, both methodological and conceptual, for work on the reception of scientific texts. Since the printed book as a source of knowledge depends upon the interpretative practices of its readers, any attempt to reconstruct the reception of a scientific text is also an attempt to reconstruct its various audiences. What geography brings to the processes by which historical audiences are recovered is an awareness that their composition and interpretive inclinations were importantly shaped by their socio-spatial location. Put simply, where Semple's readers were – both as a question of their local geographical and institutional setting, and their position within national and international networks of social exchange – mattered to how they read her book. By attending to the circulation of knowledge as a function of its distribution as a printed text (and in other representational guises) through these networks – and its reading in specific spaces – the geography of the book can help to explain both the material and epistemic spread of knowledge. This is not simply a matter of identifying where and when different audiences were exposed to that knowledge, but has to do with explaining the connections between them which constituted the networks upon which the dissemination of knowledge depended.

The plural and disparate readings of *Influences* show that the reception of Semple's ideas was not a binary defined by acceptance or rejection, but was a complicated process in which her anthropogeography was remade, and remade differently, by individual readers' encounters with it. Although the intended purpose of *Influences* was to convey Semple's ideas, it might more properly be understood to have acted as the prompt to the creation of new knowledge. The reception of *Influences* was not, however, a matter just of its reading: the circulation of Semple's anthropogeography depended upon its representation and reproduction in a number of distinct forms and different spaces. The lecture theatre and classroom were, for example, often as important as the text itself in the communication of Semple's ideas. With time, different interpretations of *Influences* became codified in lecture courses and examinations. These mediated representations mattered often as much in informing the opinion of audiences as to the particular qualities of anthropogeography than did their actual reading of the book.

The parallel between geography's engagement with environmentalism, and geographers' reading of *Influences*, illustrates the disciplining influence of Semple's book and its important function in the communication of knowledge and practice. As a number of recent studies have shown, the history of geography is also a history of geography's printed texts.[14] Attending to this common history – an approach which might be thought of as bibliographical historiography – unites biographical and prosopographical concerns in addressing the reasons why knowledge and ideas are conceived of and received differently in different places. The spatial variation in the reception of Semple's text exposes differences in the individual and collective practices of reading. These interpretative dissimilarities belie the notion of uniform institutional engagements with environmentalism, and show how the experiences and preconceptions of individual geographers, and the networks which they formed, conditioned how Semple's text functioned pedagogically in different academic venues. More particularly, it is apparent that what environmentalism was taken to be was as much the consequence of conflict and disputation as it was of concurrence and unity.

In much the same way that print culture is more than the material and technological components of print, we might think of the book as existing in more than just its textual form. *Influences* was simultaneously a printed book, and a series of representations and apprehensions. It occupied not only a textual space, but a social space too. The periodical review, examination script, scholarly lecture, academic discussion, casual conversation, and private diary were the hinterland of Semple's book – the spaces it occupied beyond its textual core, the sites where its meanings were variously created, replicated, circulated, altered, and forgotten. By attending to print's social and spatial components, the geography of the book contributes to mapping the processes by which books exist both as material objects and as cultural artefacts. In so doing, it suggests new ways in which the circulation and consumption of texts, and thus of knowledge, are understood. The work of book geography – and its exciting potential – lies precisely in its conceptualization of 'knowledge in transit', and the perspective it can bring to understanding how ideas move between places, how they are championed and challenged, accepted and repudiated.[15]

Notes

CHAPTER 1

1   The Excursion's history and itinerary is detailed in, among other sources, *Memorial Volume of the Transcontinental Excursion of 1912 of the American Geographical Society of New York* (New York: American Geographical Society, 1915); Brigham, Albert P., 'Notes on the Transcontinental Excursion of the American Geographical Society', *The Journal of Geography* 6/4 (1913), pp. 155–58; and Wright, John K., 'British geography and the American Geographical Society, 1851–1951', *The Geographical Journal* 118/2 (1952), pp. 153–67.

2   Beckit, Henry O., 'The United States National Parks', *The Geographical Journal* 42/4 (1913), pp. 333–42, p. 336.

3   Beckit: 'United States National Parks', p. 337.

4   Semple, Ellen C., *Influences of Geographic Environment on the Basis of Ratzel's System of Anthropo-geography* (New York: Henry Holt and Company, 1911).

5   Leighly, John (ed.), *Land and Life: A Selection of Writings of Carl Ortwin Sauer* (Berkeley: University of California Press, 1963), p. 5.

6   Frenkel, Stephen, 'Geography, empire, and environmental determinism', *Geographical Review* 82/2 (1992), pp. 143–53, p. 144.

7   Hartshorne, Richard, *The Nature of Geography: A Critical Survey of Current Thought in the Light of the Past* (Lancaster, PA: Association of American Geographers, 1939), p. 122.

8   Smith, Neil, 'Geography as museum: private history and conservative idealism in *The Nature of Geography*', in J. N. Entrikin and S. D. Brunn (eds), *Reflections on Richard Hartshorne's* The Nature of Geography (Washington: Association of American Geographers, 1989), pp. 91–120, p. 93.

9   Atwood, Wallace W., 'An appreciation of Ellen Churchill Semple, 1863–1932', *The Journal of Geography* 31/6 (1932), p. 267.

10 These assessment come, in order, from *The Nation*, 21 December 1911, pp. 610–11, p. 610; *Influences of Geographic Environment*, University of Oxford, Geography and the Environment Library, M 59a, p. 299; 'Review of *Influences of Geographic Environment*, by Ellen C. Semple', *The Journal of Geography* 10/1 (1911), pp. 33–34, p. 33; and *Irish Times*, 20 October 1911.

11 I borrow the term 'interpretative communities' from Fish, Stanley, *Is There a Text in This Class? The Authority of Interpretive Communities* (Cambridge, MA: Harvard University Press, 1980).

12 Secord, James A., 'Knowledge in transit', *Isis* 95/4 (2004), pp. 654–72, p. 664.

13 Febvre, Lucien and Martin, Henri-Jean, *The Coming of the Book: The Impact of Printing 1450–1800*, tr. D. Gerard (London: NLB, 1976).

14 Elements of this transition are described in, for example, Eisenstein, Elizabeth L., *The Printing Press as an Agent of Change: Communications and Cultural Transformations in Early-Modern Europe* (Cambridge: Cambridge University Press, 1979) and Johns, Adrian, *The Nature of the Book: Print and Knowledge in the Making* (Chicago and London: University of Chicago Press, 1998).

15 Price, Leah, 'Review of *In Another Country: Colonialism, Culture, and the English Novel in India*', *Victorian Studies* 45/2 (2003), pp. 333–34, p. 334.

16 MacDonald, Bertrum H. and Black, Fiona A., 'Using GIS for spatial and temporal analyses in print culture studies', *Social Science History* 24/3 (2000), pp. 505–36.

17 Livingstone, David N., 'Science, religion and the geography of reading: Sir William Whitla and the editorial staging of Isaac Newton's writings on biblical prophecy', *The British Journal for the History of Science* 36/1 (2003), pp. 27–42; Livingstone, David N., 'Science, text and space: thoughts on the geography of reading', *Transactions of the Institute of British Geographers* 30/4 (2005), pp. 391–401; and Secord, James A., *Victorian Sensation: The Extraordinary Publication, Reception, and Secret Authorship of* Vestiges of the Natural History of Creation (Chicago and London: University of Chicago Press, 2000).

18 See, for example, Keighren, Innes M., 'Bringing geography to the book: charting the reception of *Influences of Geographic Environment*', *Transactions of the Institute of British Geographers* 31/4 (2006), pp. 525–40; Livingstone: 'Science, text and space'; Mayhew, Robert J., 'The character of English geography *c.* 1660–1800: a textual approach', *Journal of Historical Geography* 24/4 (1998), pp. 385–412; Mayhew, Robert J., 'Denaturalising print, historicising text: historical geography and the history of the book', in E. A. Gagen, H. Lorimer, and A. Vasudevan (eds), *Practising the Archive: Reflections on Methods and Practice in Historical Geography* (London: Royal Geographical Society, 2007), pp. 23–36; Mayhew, Robert J., 'Materialist hermeneutics, textuality and the history of geography: print spaces in British geography, *c.* 1500–1900', *Journal of Historical Geography* 33/3 (2007), pp. 466–88; Ogborn, Miles, '*Geographia*'s pen: writing, geography and the arts of commerce, 1660–1760', *Journal of Historical Geography* 30/2 (2004), pp. 294–315; Ogborn, Miles, 'Writing travels: power, knowledge and ritual on the English East India Company's early voyages', *Transactions of the Institute of British Geographers* 27/2 (2002), pp. 155–71; Ogborn, Miles and Withers, Charles W. J., 'Travel, trade, and empire: knowing other places, 1660–1800', in C. Wall (ed.), *A Concise Companion to the Restoration and Eighteenth Century* (Oxford: Blackwell, 2005), pp. 13–35; Ogborn, Miles and Withers, Charles W. J. (eds), *Geographies of the Book* (Farnham: Ashgate, 2010); Rupke, Nicolaas A., 'Translation studies in the history of science: the example of *Vestiges*', *The British Journal for the History of Science* 33/2 (2000), pp. 209–22; and Ryan, James R., 'History and philosophy of geography: bringing geography to book, 2000–2001', *Progress in Human Geography* 27/2 (2003), pp. 195–202.

19 See, for example, Blair, Ann, 'An early modernist's perspective', *Isis* 95/3 (2004), pp. 420–30; Blair, Ann, 'Reading strategies for coping with information overload ca. 1550–1700', *Journal of the History of Ideas* 64/1 (2003), pp. 11–28; Daston, Lorraine, 'Taking note(s)', *Isis* 95/3 (2004), pp. 443–48; Frasca-Spada, Marina and Jardine, Nick (eds), *Books and the Sciences in History* (Cambridge: Cambridge University Press, 2000); Topham, Jonathan R., 'Scientific publishing and the reading of science in nineteenth-century

Britain: a historiographical survey and guide to sources', *Studies in History and Philosophy of Science* 31/4 (2000), pp. 559–612; and Topham, Jonathan R., 'A view from the industrial age', *Isis* 95/3 (2004), pp. 431–42.

20 See, for example, Glick, Thomas F. (ed.), *The Comparative Reception of Darwinism* (Austin and London: University of Texas Press, 1974); Paty, Michel, 'The scientific reception of relativity in France', in T. F. Glick (ed.), *The Comparative Reception of Relativity* (Dordrecht: D. Reidel, 1987), pp. 113–67; and Russell, Colin A., 'The reception of Newtonianism in Europe', in D. Goodman and C. A. Russell (eds), *The Rise of Scientific Europe 1500–1800* (Sevenoaks: Hodder & Stoughton, 1991), pp. 253–78.

21 Rupke, Nicolaas A., 'A geography of enlightenment: the critical reception of Alexander von Humboldt's Mexico work', in D. N. Livingstone and C. W. J. Withers (eds), *Geography and Enlightenment* (Chicago and London: University of Chicago Press, 1999), pp. 319–43.

22 Withers, Charles W. J., *Placing the Enlightenment: Thinking Geographically About the Age of Reason* (Chicago: University of Chicago Press, 2007), p. 40.

23 See, for example, Fyfe, Aileen, 'The reception of William Paley's *Natural Theology* in the University of Cambridge', *British Journal for the History of Science* 30/3 (1997), pp. 321–35; Fyfe, Aileen, *Science and Salvation: Evangelical Popular Science Publishing in Victorian Britain* (Chicago and London: University of Chicago Press, 2004); and Livingstone: 'Science, religion and the geography of reading'.

24 Livingstone, David N., *Putting Science in Its Place: Geographies of Scientific Knowledge* (Chicago and London: University of Chicago Press, 2003), p. 115.

25 Secord: *Victorian Sensation*, p. 153.

26 Gingerich, Owen, *The Book Nobody Read: In Pursuit of the Revolutions of Nicolaus Copernicus* (London: William Heinemann, 2004).

27 Gingerich, Owen, *An Annotated Census of Copernicus'* De Revolutionibus *(Nuremberg, 1543 and Basel, 1566)* (Leiden: Brill, 2002).

28 Jackson, Heather J., '"Marginal frivolities": readers' notes as evidence for the history of reading', in R. Myers, M. Harris, and G. Mandelbrote (eds), *Owners, Annotators and the Signs of Reading* (London and New Castle, DE: The British Library and Oak Knoll Press, 2005), pp. 137–51

CHAPTER 2

1 James, Preston E., Bladen, Wilford A., and Karan, Pradyumna P., 'Ellen Churchill Semple and the development of a research paradigm', in W. A. Bladen and P. P. Karan (eds), *The Evolution of Geographic Thought in America: A Kentucky Root* (Dubuque, IA: Kendall/Hunt, 1983), pp. 28–57.

2 See, for example, her chapter on 'The geography of the Civil War' in Semple, Ellen C., *American History and its Geographic Conditions* (Boston: Houghton, Mifflin and Company, 1903), 280–309.

3 These interventions include, among others, Barnett, Clive, 'Awakening the dead: who needs the history of geography?', *Transactions of the Institute of British Geographers* 20/3 (1995), pp. 417–19; Bassin, Mark, 'Studying ourselves: history and philosophy of geography', *Progress in Human Geography* 24/3 (2000), pp. 475–87; Domosh, Mona, 'Toward a feminist historiography of geography', *Transactions of the Institute of British Geographers* 16/1 (1991), pp. 95–104; Keighren, Innes M., 'Breakfasting with William Morris Davis: everyday episodes in the history of geography', in E. A. Gagen, H. Lorimer, and A. Vasudevan (eds), *Practising the Archive: Reflections on Methods and Practice in Historical Geography* (London: Royal Geographical Society, 2007),

pp. 47–55; Livingstone, David N., *The Geographical Tradition: Episodes in the History of a Contested Enterprise* (Oxford: Blackwell, 1992); Livingstone, David N., 'The history of science and the history of geography: interactions and implications', *History of Science* 22/3 (1984), pp. 271–302; and Withers, Charles W. J. and Mayhew, Robert J., 'Rethinking "disciplinary" history: geography in British universities, *c.* 1580–1887', *Transactions of the Institute of British Geographers* 27/1 (2002), pp. 11–29.

4   Mikesell, Marvin W., 'Continuity and change', in B. W. Blouet (ed.), *The Origins of Academic Geography in the United States* (Hamden, CT: Archon Books, 1981), pp. 1–15, p. 8.

5   Hankins, Thomas L., 'In defence of biography: the use of biography in the history of science', *History of Science* 17/1 (1979), pp. 1–16, p. 5.

6   Harrison, Lowell H. and Klotter, James C., *A New History of Kentucky* (Lexington: University Press of Kentucky, 1997).

7   Bushong, Allen D., 'Ellen Churchill Semple 1863–1932', in T. W. Freeman (ed.), *Geographers: Biobibliographical Studies*, vol. 8 (London: Mansell, 1984), pp. 87–94, p. 87.

8   The geographical position of Louisville, and its historical and political implications, are discussed in Kleber, John E. (ed.), *The Encyclopedia of Louisville* (Lexington: University Press of Kentucky, 2000); Semple, Ellen C., 'Louisville, a study in economic geography', *Journal of School Geography* 4 (1900), pp. 361–70; and Wheeler, James O. and Brunn, Stanley D., 'An urban geographer before his time: C. Warren Thornthwaite's 1930 doctoral dissertation', *Progress in Human Geography* 26/4 (2002), pp. 463–86.

9   Colby, Charles C., 'Ellen Churchill Semple', *Annals of the Association of American Geographers* 23/4 (1933), pp. 229–40, p. 229.

10 Bingham, Millicent T., 'Ellen Churchill Semple, geographer', *Vassar Quarterly*, July (1932), unpaginated.

11 Colby: 'Ellen Churchill Semple', p. 229.

12 James, Bladen, and Karan: 'Ellen Churchill Semple', p. 29.

13 *The Evening Post*, 9 November 1912.

14 Aspects of Semple's experience at Vassar are described in Bingham: 'Ellen Churchill Semple'; Lewis, Carolyn B., 'The biography of a neglected classic: Ellen Churchill Semple's *The Geography of the Mediterranean Region*' (Ph.D. diss., University of South Carolina, 1979).

15 *The Evening Post*, 9 November 1912.

16 Bronson, Judith C., 'Ellen Semple: contributions to the history of American geography' (Ph.D. diss., Saint Louis University, 1973).

17 University of Kentucky, Special Collections and Digital Programs (hereafter UK), 46M139, Box 2. Baugh, Ruth E., 'Ellen Churchill Semple, the great lady of American geography', 1961, p. 3.

18 *Bulletin of Vassar College: Alumnae Biographical Register Issue* (Poughkeepsie, NY: Vassar College, 1939).

19 Bushong: 'Ellen Churchill Semple', p. 88.

20 Johnson, Joan M., *Southern Women at the Seven Sister Colleges: Feminist Values and Social Activism, 1875–1915* (Athens and London: University of Georgia Press, 2008).

21 Brandeis, Adele, 'Ellen Semple as I remember her', *Courier Journal*, 8 January 1963, unpaginated.

22 Brandeis: 'Ellen Semple', unpaginated.

23 Harvard University Archives (hereafter HUA), HUG 4877.410. Charles C. Colby to Whittlesey, 18 June 1932.

24 Colby: 'Ellen Churchill Semple', p. 231.

25 HUA, HUG 4877.410. Charles C. Colby to Whittlesey, 18 June 1932.

26 *The Evening Post*, 9 November 1912.

27 University of California at Berkeley, The Bancroft Library (hereafter UCB), CU-468, Box 1, Folder 2. Glacken to Thomas R. Smith, 19 April 1963.

28 *The Evening Post*, 9 November 1912.

29 Royal Geographical Society (hereafter RGS), Correspondence Block 1911–1920. Semple to John S. Keltie, 30 October 1912.

30 Clark University, Archives and Special Collections (hereafter CU), B4-18-11. Duren J. H. Ward to Atwood, 2 July 1932.

31 James, Bladen, and Karan: 'Ellen Churchill Semple', p. 30.

32 UK, 46M139, Box 2. Ruth E., 'Ellen Churchill Semple, the great lady of American geography', 1961, p. 3.

33 RGS, Correspondence Block 1911–1920. Semple to John S. Keltie, 30 October 1912.

34 CU, B4-18-11. Duren J. H. Ward to Atwood, 2 July 1932.

35 *The Evening Post*, 9 November 1912.

36 Luxenberg, Adele, 'Women at Leipzig', *The Nation*, 4 October 1894, pp. 247–48.

37 Buttmann, Günther, *Friedrich Ratzel: Leben und Werk eines deutschen Geographen 1844–1904* (Stuttgart: Wissenschaftliche Verlagsgesellschaft, 1977).

38 RGS, Correspondence Block 1911–1920. Semple to John S. Keltie, 30 October 1912.

39 RGS, Correspondence Block 1911–1920. Semple to John S. Keltie, 30 October 1912.

40 RGS, Correspondence Block 1911–1920. Semple to John S. Keltie, 30 October 1912.

41 American Philosophical Society Library (hereafter APS), B Sm59. 'Journal of European trips, 1901–1903', 27 November 1901, unpaginated.

42 APS, B Sm59. 'Journal of European trips, 1901–1903', 9 February 1902, unpaginated.

43 Bingham: 'Ellen Churchill Semple', unpaginated.

44 RGS, Correspondence Block 1911–1920. Semple to John S. Keltie, 30 October 1912.

45 This recollection can be found in the author's handwritten postscript to a copy of *Influences of Geographic Environment* held at the American Geographical Society Library (call number GF31 .S5).

46 The contours of nineteenth-century German geography are traced by, among other scholars, Keltie, John S. and Howarth, Osbert J. R., *History of Geography* (London: Watts, 1913); Mackinder, Halford J., 'Modern geography, German and English', *The Geographical Journal* 6/4 (1895), pp. 367–79; Martin, Geoffrey J., *All Possible Worlds: A History of Geographical Ideas*, 4th ed. (Oxford: Oxford University Press, 2005); and Tang, Chenxi, *The Geographical Imagination of Modernity: Geography, Literature, and Philosophy in German Romanticism* (Palo Alto, CA: Stanford University Press, 2008).

47 Keltie and Howarth: *History of Geography*, p. 144.

48 Sachs, Aaron J., *The Humboldt Current: Nineteenth-Century Exploration and the Roots of American Environmentalism* (New York: Viking, 2006).

49 Tatham, George, 'Environmentalism and possibilism', in G. Taylor (ed.), *Geography in the Twentieth Century* (London: Methuen, 1957), pp. 128–62, p. 128.

50 Elements of the history of environmentalist thought are provided by, for example, Glacken, Clarence J., *Traces on the Rhodian Shore: Nature and Culture in Western Thought From Ancient Times to the End of the Eighteenth Century*

(Berkeley and Los Angeles: University of California Press, 1967); Tatham: 'Environmentalism and possibilism'; and Withers, Charles W. J., *Placing the Enlightenment: Thinking Geographically About the Age of Reason* (Chicago: University of Chicago Press, 2007).

51 Broek, Jan O. M., *Geography: Its Scope and Spirit* (Columbus, OH: Charles E. Merrill Books, 1965), p. 15.

52 Tatham, George, 'Geography in the nineteenth century', in G. Taylor (ed.), *Geography in the Twentieth Century* (London: Methuen, 1957), pp. 28–69.

53 Gage, William L., *The Life of Carl Ritter: Late Professor of Geography in the University of Berlin* (New York: Charles Scribner and Co, 1867).

54 Sauer, Carl O., 'Recent developments in cultural geography', in E. C. Hayes (ed.), *Recent Developments in the Social Sciences* (Philadelphia: J. B. Lippincott, 1927), pp. 154–212, p. 166.

55 Bassin: 'Studying ourselves': p. 482.

56 Koelsch, William A., 'Franz Boas, geographer, and the problem of disciplinary identity', *Journal of the History of the Behavioural Sciences* 40/1 (2004), pp. 1–22, p. 3.

57 Broek, Jan O. M., *Compass of Geography* (Columbus: Charles E. Merril Books, 1966).

58 Tatham: 'Geography in the nineteenth century', p. 58.

59 Schelhaas, Bruno and Hönsch, Ingrid, 'History of German geography: worldwide reputation and strategies of nationalism and institutionalisation', in G. S. Dunbar (ed.), *Geography: Discipline, Profession and Subject Since 1870* (Dordrecht: Kluwer, 2001), pp. 9–44, p. 16.

60 Hunter, James M., *Perspectives on Ratzel's Political Geography* (Lanham, MD: University Press of America, 1983).

61 Aspects of Richthofen's biography are detailed in Kolb, Albert, 'Ferdinand Freiherr von Richthofen 1833–1905', in T. W. Freeman (ed.), *Geographers: Biobibliographical Studies*, vol. 7 (London: Mansell, 1983), pp. 109–15; and Ravenstein, E. G., 'Obituary: Ferdinand Freiherr von Richthofen', *The Geographical Journal* 26/6 (1905), pp. 679–82.

62 Crone, Gerald R., *Modern Geographers: An Outline of Progress in Geography Since 1800 A.D.* (London: Royal Geographical Society, 1951), p. 35.

63 Martin: *All Possible Worlds*, p. 167.

64 Dickinson, Robert E., *The Makers of Modern Geography* (New York: Frederick A. Praeger, 1969); Hunter: *Ratzel's political geography*; and Sanguin, André-Louis, 'En relisant Ratzel', *Annales de Géographie* 555 (1990), pp. 579–94.

65 Sauer, Carl O., 'The formative years of Ratzel in the United States', *Annals of the Association of American Geographers* 61/2 (1971), pp. 245–54, p. 245.

66 Ratzel, Friedrich, *Sein und Werden der organischen Welt. Eine populäre Schöpfungsgeschichte* (Leipzig: Gebhart und Reisland, 1869); and Wanklyn, Harriet, *Friedrich Ratzel: A Biographical Memoir and Bibliography* (Cambridge: Cambridge University Press, 1961).

67 Bassin, Mark, 'Friedrich Ratzel 1844–1904', in T. W. Freeman (ed.), *Geographers: Biobibliographical Studies*, vol. 11 (London: Mansell, 1987), pp. 123–32.

68 Beck, Hanno, 'Moritz Wagner als geograph', *Erdkunde* 7 (1953), pp. 125–28.

69 Smith, Woodruff D., 'Friedrich Ratzel and the origins of Lebensraum', *German Studies Review* 3/1 (1980), pp. 51–68.

70 Campbell, John A. and Livingstone, David N., 'Neo-Lamarckism and the development of geography in the United States and Great Britain', *Transactions of the Institute of British Geographers* 8/3 (1983), pp. 267–94.

71 Wanklyn: *Friedrich Ratzel*, p. 11.

72   Bassin, Mark, 'Reply: reductionism or redux? or the convolutions of con-
      textualism', *Annals of the Association of American Geographers* 83/1 (1993), pp.
      163–66; Natter, Wolfgang, 'Friedrich Ratzel's spatial turn: identities of dis-
      ciplinary space and its borders between the anthropo- and political geog-
      raphy of Germany and the United States', in H. van Houtum, O. Kramsch,
      and Z. Wolfgang (eds), *Bordering Space* (Aldershot: Ashgate, 2005), pp. 171–88.
73   Bassin, Mark, 'History and philosophy of geography', *Progress in Human Geog-
      raphy* 21/4 (1997), pp. 563–72, p. 568.
74   Mikesell, Marvin W., 'Ratzel, Friedrich', in D. L. Sills (ed.), *International En-
      cyclopedia of Social Sciences*, vol. 13 (New York: MacMillan and Free Press,
      1968), pp. 327–29.
75   UCB, CU-468, Box 7, Folder 34. Mss. on ideas, 1974, unpaginated.
76   UCB, CU-468, Box 7, Folder 34. Mss. on ideas, 1974, unpaginated.
77   Aspects of Spencer's perspective are outlined by Livingstone: *Geographical
      Tradition*; Peet, Richard, 'The social origins of environmental determinism',
      Annals of the Association of American Geographers 75/3 (1985), pp. 309–33;
      Shapin, Steven, 'Man with a plan: Herbert Spencer's theory of everything',
      The New Yorker, 13 August 2007, pp. 75–79; and Simon, Walter M., 'Herbert
      Spencer and the "social organism"', *Journal of the History of Ideas* 21/2 (1960),
      pp. 294–99.
78   Tatham: 'Geography in the nineteenth century', p. 63.
79   Stoddart, David R., 'Darwin's impact on geography', *Annals of the Association
      of American Geographers* 56/4 (1966), pp. 683–98.
80   UCB, CU-468, Box 1, Folder 2. Glacken to Thomas R. Smith, 19 April 1963.
81   Tatham: 'Geography in the nineteenth century', p. 64.
82   Tatham: 'Geography in the nineteenth century', p. 64.
83   Weikart, Richard, 'The origins of Social Darwinism in Germany, 1859–1895',
      Journal of the History of Ideas 54/3 (1993), pp. 469–88.
84   Crone: *Modern Geographers*, p. 37.
85   APS, Mss.B.B61, Series IV. 'Franz Boas: his work as described by some of his
      contemporaries', unpaginated.
86   Koelsch: 'Franz Boas', p. 1.
87   Koelsch: 'Franz Boas', p. 5.
88   Quoted in Koelsch: 'Franz Boas', p. 6.
89   APS, Mss.B.B61, Series IV. 'Franz Boas: his work as described by some of his
      contemporaries', unpaginated.
90   APS, Mss.B.B61, Series IV. 'Franz Boas: his work as described by some of his
      contemporaries', unpaginated.
91   Boas, Franz, 'The study of geography', *Science* 9/210 (1887), pp. 137–41;
      Koelsch, William A., *Clark University, 1887–1987: A Narrative History* (Worces-
      ter, MA: Clark University Press, 1987).
92   RGS, Correspondence Block 1921–1930. Semple to Hugh R. Mill, 19 Decem-
      ber 1924.
93   Harris, Chauncy D. and Fellmann, Jerome D., 'Geographical serials', *Geo-
      graphical Review* 40/4 (1950), pp. 649–56.
94   RGS, Correspondence Block 1921–1930, Annual Awards 1922, Paper propos-
      ing Ellen Churchill Semple, A.M., LL.D., unpaginated.
95   Cahnman, Werner J., 'Methods of geopolitics', *Social Forces* 21/2 (1942), pp.
      147–54.
96   Buttimer, Anne, *Society and Milieu in the French Geographic Tradition* (Chicago:
      Rand McNally, 1971).
97   Semple, Ellen C., 'Review of *Anthropogeographie* by Friedrich Ratzel', *Annals of
      the American Academy of Political and Social Science* 16 (1900), pp. 137–39.

187

98 Wanklyn: *Friedrich Ratzel*, p. 23.
99 UK, 46M139, Box 2. Baugh, Ruth E., 'Ellen Churchill Semple, the great lady of American geography', 1961, p. 5.
100 RGS, Correspondence Block 1921–1930. Semple to Hugh R. Mill, 19 December 1924.
101 *Review of Reviews*, July 1911.
102 Bronson: 'Ellen Semple', p. 90.
103 James, Preston E. and Martin, Geoffrey J., *All Possible Worlds: A History of Geographical Ideas*, 2nd ed. (New York: John Wiley & Sons, 1981), pp. 304–05.
104 James, Edward T., James, Janet W., and Boyer, Paul S. (eds), *Notable American Women 1607–1950. A Biographical Dictionary*, vol. 3 (Cambridge, MA: Belknap Press, 1971), p. 260.
105 *The Courier-Journal*, 13 June 1903; Thompson, Lawrence S., 'Alice Caldwell Hegan Rice', in R. Bain, J. M. Flora, and L. D. Rubin (eds), *Southern Writers: A Biographical Dictionary* (1979), pp. 381–82.
106 Ratzel, Friedrich, 'Studies in political areas. The political territory in relation to earth and continent', *The American Journal of Sociology* 3/3 (1897), pp. 297–313; Ratzel, Friedrich, 'Studies in political areas. II. Intellectual, political, and economic effects of large areas', *The American Journal of Sociology* 3/4 (1898), p. 449–63; Ratzel, Friedrich, 'Studies in political areas. III. The small political area', *The American Journal of Sociology* 4/3 (1898), pp. 366–79; Semple, Ellen C., 'Review of *Politische geographie der vereinigten staaten von Amerika*, by Friedrich Ratzel', *Annals of the American Academy of Political and Social Science* 4 (1894), pp. 139–40; and Semple, Ellen C., 'Review of *Die staat und sein boden*, by Friedrich Ratzel', *Annals of the American Academy of Political and Social Science* 9 (1897), pp. 102–04.
107 Croly, Jane C., *The History of the Woman's Club Movement in America* (New York: Henry G. Allen & Co., 1898), p. 173.
108 Dryer, Charles R., 'A century of geographic education in the United States', *Annals of the Association of American Geographers* 14/3 (1924), pp. 117–49, p. 148; Visher, Stephen S., 'Richard Elwood Dodge, 1868–1952', *Annals of the Association of American Geographers* 42/4 (1952), pp. 318–21.
109 Semple, Ellen C., 'The influence of the Appalachian barrier upon colonial history', *Journal of School Geography* 1 (1897), pp. 33–41.
110 Semple: *American History*, p. 36.
111 Gelfand, Lawrence E., 'Ellen Churchill Semple: her geographical approach to American history', *The Journal of Geography* 53/1 (1954), pp. 30–37, p. 30.
112 Bogue, Allan G., '"Not by bread alone": the emergence of the Wisconsin idea and the departure of Frederick Jackson Turner', *Wisconsin Magazine of History* 2002, pp. 10–23, p. 13.
113 Quoted in Benson, Lee, *Turner and Beard: American Historical Writing Reconsidered* (Glencoe, IL: The Free Press, 1960), p. 23.
114 For a counterpoint to Turner's conception of individualism, see Boatright, Mody C., 'The myth of frontier individualism', *Southwestern Social Science Quarterly* 22 (1941), pp. 14–32.
115 Campbell and Livingstone: 'Neo-Lamarckism'; Kearns, Gerry, 'Closed space and political practice: Frederick Jackson Turner and Halford Mackinder', *Environment and Planning D: Society and Space* 2/1 (1984), pp. 23–34.
116 Livingstone, David N., 'Environment and inheritance: Nathaniel Southgate Shaler and the American frontier', in B. W. Blouet (ed.), *The Origins of Academic Geography in the United States* (Hamden, CT: Archon Books, 1981), pp. 123–38, p. 125.

117 Block, Robert H., 'Frederick Jackson Turner and American geography', *Annals of the Association of American Geographers* 70/1 (1980), pp. 31–42, p. 32.
118 Brewer, William M., 'The historiography of Frederick Jackson Turner', *The Journal of Negro History* 44/3 (1959), pp. 240–59, p. 240.
119 Coleman, William, 'Science and symbol in the Turner frontier hypothesis', *The American Historical Review* 72/1 (1966), pp. 22–49.
120 Block: 'Frederick Jackson Turner'.
121 Benson, Lee, 'Achille Loria's influence on American economic thought: including his contributions to the frontier hypothesis', *Agricultural History* 24/4 (1950), pp. 182–99.
122 Meadows, Paul, 'Achille Loria: agrarian determinist', *American Journal of Economics and Sociology* 10/2 (1951), pp. 175–54.
123 Freund, Rudolf, 'Turner's theory of social evolution', *Agricultural History* 19/2 (1945), 78–87.
124 Coleman: 'Science and symbol', p. 24.
125 Block, 'Frederick Jackson Turner', p. 31.
126 American Geographical Society Library (hereafter AGSL). Clarence J. Glacken to John K. Wright, 11 May 1961.
127 Bladen, Wilford A., 'Nathaniel Southgate Shaler and early American geography', in W. A. Bladen and P. P. Karan (eds), *The Evolution of Geographic Thought in America: A Kentucky Root* (Dubuque, IA: Kendall/Hunt, 1983), pp. 13–27.
128 Shaler, Nathaniel S., *Nature and Man in America* (New York: Charles Scribner's Sons, 1891).
129 Shaler: *Nature and Man*, p. 360.
130 Shaler: *Nature and Man*, p. 360.
131 Livingstone, David N., *Nathaniel Southgate Shaler and the Culture of American Science* (Tuscaloosa: University of Alabama Press, 1987), p. 135.
132 Livingstone: 'Environment and inheritance', p. 130.
133 Livingstone: 'Environment and inheritance', p. 130.
134 Livingstone: 'Environment and inheritance', p. 132.
135 For a detailed discussion of the history of monogenist and polygenist philosophies, see, for example, Livingstone, David N., *Adam's Ancestors: Race, Religion, and the Politics of Human Origins* (Baltimore: Johns Hopkins University Press, 2008).
136 Shaler: *Nature and Man*, p. 365.
137 Marsh, George P., *Man and Nature; or, Physical Geography as Modified by Human Action* (New York: Charles Scribner, 1864).
138 Clark, Brett and Foster, John B., 'George Perkins Marsh and the transformation of the earth: an introduction to Marsh's *Man and Nature*', *Organization & Environment* 15/2 (2002), pp. 164–69.
139 Elder, John, *Pilgrimage to Vallombrosa: From Vermont to Italy in the Footsteps of George Perkins Marsh* (Charlottesville: University of Virginia Press, 2006).
140 Koelsch, William A., 'Seedbed of reform: Arnold Guyot and school geography in Massachusetts, 1849–1855', *The Journal of Geography* 107/2 (2008), pp. 35–42; Koelsch, William A., 'Three friends of Swiss-American science: Louis Agassiz, Arnold Guyot, and Cornelius C. Felton', *Swiss American Historical Society Review* 44/1 (2008), pp. 45–59; and Lowenthal, David, *George Perkins Marsh: Prophet of Conservation* (Seattle: University of Washington Press, 2000).
141 Martin, Geoffrey J., 'The emergence and development of geographic thought in New England', *Economic Geography* 74 (1998), pp. 1–13.

142 Guyot, Arnold H., *Earth and Man: Lectures on Comparative Physical Geography in its Relation to the History of Mankind* (Boston: Gould, Kendall and Lincoln, 1849).

143 Quoted in Lowenthal: *George Perkins Marsh: Prophet of Conservation*, p. 267.

144 Lowenthal: *George Perkins Marsh: Prophet of Conservation*, p. 302.

145 Walls, Laura D., *The Passage to Cosmos: Alexander von Humboldt and the Shaping of America* (Chicago and London: University of Chicago Press, 2009).

146 Semple, Ellen C., 'A comparative study of the Atlantic and Pacific oceans. Part I', *Journal of School Geography* 3/4 (1899), pp. 121–29; Semple, Ellen C., 'A comparative study of the Atlantic and Pacific oceans. Part II', *Journal of School Geography* 3/5 (1899), pp. 172–80; Semple, Ellen C., 'The development of Hanse towns in relation to their geographical environment', *Journal of the American Geographical Society of New York* 31/3 (1899), pp. 236–55; Semple, Ellen C., 'The Indians of southeastern Alaska in relation to their environment', *Journal of School Geography* 2/6 (1898), pp. 206–15; Semple: 'Louisville'; and Semple, Ellen C., 'Some geographic causes determining the location of cities', *Journal of School Geography* 2 (1897), pp. 206–31.

147 Dodge, Richard E., 'The social function of geography', *Journal of School Geography* 2/9 (1898), pp. 328–36, p. 335.

148 Merrill, James A., 'A suggestive course in geography', *Journal of School Geography* 2/9 (1898), pp. 321–28, p. 321.

149 HUA, HUG 4877.417. Whittlesey to Charles C. Colby, 26 January 1954.

150 Semple, Ellen C., 'A new departure in social settlements', *Annals of the American Academy of Political and Social Science* 15 (1900), pp. 157–60, p. 158.

151 Semple: 'New departure in social settlements', p. 158. Italicization in original.

152 Semple: 'New departure in social settlements', p. 158.

153 Semple: 'New departure in social settlements', p. 158.

154 Shapiro, Henry D., *Appalachia on Our Mind: The Southern Mountains and Mountaineers in the American Consciousness, 1870–1920* (Chapel Hill: University of North Carolina Press, 1986).

155 Daingerfield, Henderson, 'Social settlement and educational work in the Kentucky mountains', *Journal of Social Science* 39 (1901), pp. 176–89, p. 178.

156 Wilson, Shannon H., *Berea College: An Illustrated History* (Lexington: University Press of Kentucky, 2006), p. 16.

157 Semple: 'New departure in social settlements', p. 158.

158 Semple, Ellen C., 'The Anglo-Saxons of the Kentucky Mountains: a study in anthropogeography', *The Geographical Journal* 17/6 (1901), pp. 561–94.

159 Colby: 'Ellen Churchill Semple', p. 232.

160 Semple: 'Anglo-Saxons of the Kentucky Mountains', p. 588.

161 Semple: 'Anglo-Saxons of the Kentucky Mountains', p. 591.

162 Semple: 'Anglo-Saxons of the Kentucky Mountains', p. 592.

163 Semple: 'Anglo-Saxons of the Kentucky Mountains', p. 593.

164 Semple: 'Anglo-Saxons of the Kentucky Mountains', p. 594.

165 Semple: 'Anglo-Saxons of the Kentucky Mountains', p. 594.

166 Semple: 'Anglo-Saxons of the Kentucky Mountains', p. 598.

167 Semple: 'Anglo-Saxons of the Kentucky Mountains', p. 597.

168 Semple: 'Anglo-Saxons of the Kentucky Mountains', p. 597.

169 Beckinsale, Robert P., 'W. M. Davis and American geography: 1880–1934', in B. W. Blouet (ed.), *The Origins of Academic Geography in the United States* (Hamden, CT: Archon Books, 1981), pp. 107–22.

170 Davis, William M., 'Current notes of physiography', *Science* 14/351 (1901), pp. 457–59.

171 Colby: 'Ellen Churchill Semple', p. 232.

172 *The Courier-Journal*, 11 January 1903.

173 *The Courier-Journal*, 11 January 1903.

174 *The Courier-Journal*, 13 June 1903.

175 UCB, BANC MSS 77/170 c, Box 12. Preston E. James to Sauer, 25 January 1972.

176 James, Preston E., 'Albert Perry Brigham 1855–1932', in T. W. Freeman and P. Pinchemel (eds), *Geographers: Biobibliographical Studies*, vol. 2 (London: Mansell, 1978), pp. 13–17.

177 Hart, Albert B., 'Review of *Geographic Influences in American History*, by Albert P. Brigham and *American History and its Geographic Conditions*, by Ellen C. Semple', *The American Historical Review* 9/3 (1904), pp. 571–72, p. 571.

178 Keasby, Lindley M., 'Review of *American History and its Geographic Conditions*, by Ellen C. Semple and *Geographic Influences in American History* by Albert P. Brigham', *Political Science Quarterly* 19/3 (1904), pp. 501–2; and Turner, Frederick J., 'Geographical interpretations of American history. Review of *American History and its Geographic Conditions*, by Ellen C. Semple and *Geographic Influences in American History*, by Albert P. Brigham', *The Journal of Geography* 4/1 (1905), pp. 34–37, p. 36.

179 Herbertson, Andrew J., 'Two books on the historical geography of the United States. Review of *Geographic Influences in American History*, by Albert P. Brigham and *American History and its Geographic Conditions*, by Ellen C. Semple', *The Geographical Journal* 23/5 (1904), pp. 674–77; and Tarr, Ralph S., 'Review of *American History and its Geographic Conditions*, by Ellen C. Semple', *Bulletin of the American Geographical Society* 35/5 (1903), pp. 566–70.

180 Herbertson, Andrew J., 'Geography in the university', *Scottish Geographical Magazine* 18/3 (1902), pp. 124–32.

181 James, Preston E., 'Geographical ideas in America, 1890–1914', in B. W. Blouet (ed.), T*he Origins of Academic Geography in the United States* (Hamden, CT: Archon Books, 1981), pp. 319–26.

182 See, for example, *The Evening Post*, 9 November 1912; RGS, Correspondence Block 1911–1920. Semple to John S. Keltie, 30 October 1912; and Price, Edward T., 'Geography at the University of Oregon', *Association of Pacific Coast Geographers Yearbook* 52 (1990), pp. 140–52.

183 RGS, Correspondence Block 1921–1930, Annual Awards 1922, Paper proposing Ellen Churchill Semple, A.M., LL.D., unpaginated.

184 Barton, Thomas F. and Karan, Pradyumna P., *Leaders in American Geography*, vol. 1 (Mesilla: The New Mexico Geographical Society, 1992), p. 81.

185 *The Courier-Journal*, 13 June 1903.

186 Quoted in Hunter: *Ratzel's Political Geography*, p. 531.

187 Genthe, Martha K. and Semple, Ellen C., 'Tributes to Friedrich Ratzel', *Bulletin of the American Geographical Society* 36/9 (1904), pp. 550–53.

188 Semple, Ellen C., 'Emphasis upon anthropo-geography in schools', *The Journal of Geography* 3/8 (1904), pp. 366–74.

189 James, Preston E. and Martin, Geoffrey J., *The Association of American Geographers, the First Seventy-Five Years, 1904–1979* (Washington, DC: The Association of American Geographers, 1978), p. 31.

190 James and Martin: *Association of American Geographers*, p. 31.

191 Dunbar, Gary S., 'Credentialism and careerism in American geography, 1890–1915', in B. W. Blouet (ed.), *The Origins of Academic Geography in the United States* (Hamden, CT: Archon Books, 1981), pp. 71–88; and Wright, John K., *Geography in the Making: The American Geographical Society 1851–1951* (New York: American Geographical Society, 1952).

192 Lowenthal, David, 'Fruitful liaison or folie à deux? The AAG and AGS', *The Professional Geographer* 57/3 (2005), pp. 468–73, p. 468.

193 James, Preston E. and Ehrenberg, Ralph, 'The original members of the Association of American Geographers', *The Professional Geographer* 27/3 (1975), pp. 327–34.

194 Monk, Janice J., 'Women, gender, and the histories of American geography', *Annals of the Association of American Geographers* 94/1 (2004), pp. 1–22, p. 2.

195 James and Martin: *Association of American Geographers*, p. 31.

196 See, for example, Berman, Mildred, 'Sex discrimination and geography: the case of Ellen Churchill Semple', *The Professional Geographer* 26/1 (1974), pp. 8–11; Bronson, Judith C., 'A further note on sex discrimination and geography: the case of Ellen Churchill Semple', *The Professional Geographer* 27/1 (1975), pp. 111–12; Monk: 'Histories of American geography'; and Monk, Janice J., 'Women's worlds at the American Geographical Society', *Geographical Review* 93/2 (2004), pp. 237–57.

197 See James and Ehrenberg: 'Original members of the Association of American Geographers'; and James, Preston E. and Martin, Geoffrey J., 'On AAG history', *The Professional Geographer* 31/4 (1979), pp. 353–57.

198 RGS, Correspondence Block 1881–1910. Semple to John S. Keltie, 12 August 1905.

199 RGS, Correspondence Block 1881–1910. Semple to John S. Keltie, 2 September 1905.

200 Pattison, William D., 'Rollin Salisbury and the establishment of geography at the University of Chicago', in B. W. Blouet (ed.), *The Origins of Academic Geography in the United States* (Hamden, CT: Archon Books, 1981), pp. 151–63.

201 Barrows, Harlan H., 'The department of geography: thirty years ago and now', *The University Record* 19/3 (1933), pp. 197–99, p. 198.

202 Schneider, Allan F., 'Chamberlin, Salisbury, and Collie: a tale of three Beloit College geologists', *Geoscience Wisconsin* 18 (2001), pp. 9–20.

203 James and Martin: *All Possible Worlds*, p. 311.

204 Pattison, William D., 'Rollin D. Salisbury 1858–1922', in T. W. Freeman (ed.), *Geographers: Biobibliographical Studies*, vol. 6 (London: Mansell, 1982), p. 107.

205 Pattison: 'Rollin D. Salisbury', p. 107.

206 UC, Robert S. Platt Papers, Box 3, Folder 6. Platt to Walter M. Kollmorgen, 12 May 1956.

207 UC, Robert S. Platt Papers, Box 3, Folder 6. Platt to Walter M. Kollmorgen, 12 May 1956.

208 James and Martin: *All Possible Worlds*, p. 311.

209 Foster, Alice, 'The new department in its setting', in *A Half Century of Geography – What Next?* (Chicago: Department of Geography, University of Chicago, 1955), pp. 1–7.

210 UC, Robert S. Platt Papers, Box 3, Folder 6. Platt to Walter M. Kollmorgen, 12 May 1956.

211 UC, Robert S. Platt Papers, Box 3, Folder 6. Platt to Walter M. Kollmorgen, 12 May 1956.

212 Grossman, Lary, 'Man-environment relationships in anthropology and geography', *Annals of the Association of American Geographers* 67/1 (1977), pp. 126–44, p. 128.

213 Barrows, Harlan H., *Lectures on the Historical Geography of the United States as Given in 1933*, W. A. Koelsch (ed.) (Chicago: Department of Geography, University of Chicago, 1962).

214 Barrows, Harlan H., 'Geography as human ecology', *Annals of the Association of American Geographers* 13/1 (1923), pp. 1–14; and Koelsch, William A., 'The

historical geography of Harlan H. Barrows', *Annals of the Association of American Geographers* 59/4 (1969), pp. 632–51.

215 Koelsch: 'Historical geography of Harlan H. Barrows', p. 633.
216 Bushong: 'Ellen Churchill Semple', p. 89.
217 UC, Robert S. Platt Papers, Box 10, Folder 16. 'Changes in geographic thought', 19 March 1950, unpaginated.
218 *Circular of Information for the Year 1906–1907* (Chicago: University of Chicago Press, 1906), p. 122.
219 UC, Robert S. Platt Papers, Box 10, Folder 16. 'Changes in geographic thought', 19 March 1950, unpaginated.
220 James, Bladen, and Karan: 'Ellen Churchill Semple', p. 33.
221 Bushong, Allen D., 'Geographers and their mentors: a genealogical view of American academic geography', in B. W. Blouet (ed.), *The Origins of Academic Geography in the United States* (Hamden, CT: Archon Books, 1981), pp. 193–219.
222 *The Evening Post*, 9 November 1912.
223 RGS, Correspondence Block 1911–1920. Semple to John S. Keltie, 2 April 1911.
224 Hawley, Arthur J., 'Environmental perception: nature and Ellen Churchill Semple', *The Southeastern Geographer* 8 (1968), pp. 54–59.
225 Semple: *Influences of Geographic Environment*, p. vii.
226 Barnes, Harry E., 'Some contributions of sociology to modern political theory', *The American Political Science Review* 15/4 (1921), pp. 487–533, p. 505.
227 Her paper was published as Semple, Ellen C., 'Geographical boundaries. Part I', *Bulletin of the American Geographical Society* 39/7 (1907), pp. 385–97; and Semple, Ellen C., 'Geographical boundaries. Part II', *Bulletin of the American Geographical Society* 39/8 (1907), pp. 449–63.
228 Block: 'Frederick Jackson Turner', p. 38.
229 'The meeting of the American Historical Association at Madison', *The American Historical Review* 13/3 (1908), pp. 433–58, p. 436.
230 'Meeting of the American Historical Association', p. 437.
231 Quoted in Block: 'Frederick Jackson Turner', p. 38.
232 'Meeting of the American Historical Association', p. 437.
233 AGSL. Charles C. Colby to John K. Wright, 23 May 1961.
234 RGS, Correspondence Block 1881–1910. Semple to John S. Keltie, 21 April 1907.
235 RGS, Correspondence Block 1881–1910. Semple to John S. Keltie, 21 April 1907. Underlining in original.
236 RGS, Correspondence Block 1881–1910. Semple to John S, Keltie, 8 March 1908.
237 Semple, Ellen C., 'Geographical location as a factor in history', *Bulletin of the American Geographical Society* 40/2 (1908), pp. 65–81; Semple, Ellen C., 'Oceans and enclosed seas: a study in anthropo-geography', *Bulletin of the American Geographical Society* 40/4 (1908), pp. 193–209; and Semple, Ellen C., 'The operation of geographic factors in history', *Bulletin of the American Geographical Society* 41/7 (1909), pp. 422–39.
238 Semple: 'Operation of geographic factors in history', p. 65.
239 Semple, Ellen C., 'The Anglo-Saxons of the Kentucky Mountains: a study in anthropogeography', *Bulletin of the American Geographical Society* 42/8 (1910), pp. 588–623.
240 RGS, Correspondence Block 1881–1910. Semple to John S. Keltie, 6 March 1910.
241 See, for example, Semple, Ellen C., 'Coast peoples. Part I', *The Geographical Journal* 31/1 (1908), pp. 72–90; and Semple, Ellen C., 'Coast peoples. Part II', *The Geographical Journal* 31/2 (1908), pp. 170–87.

242 RGS, Correspondence Block 1881–1910. Semple to John S. Keltie, 6 March 1910.
243 RGS, Correspondence Block 1881–1910. Semple to John S. Keltie, 6 March 1910.
244 RGS, Correspondence Block 1881–1910. John S. Keltie to Semple, 16 March 1910.
245 RGS, Correspondence Block 1911–1920. Semple to John S. Keltie, 2 April 1911.
246 RGS, Correspondence Block 1911–1920. Semple to John S. Keltie, 2 April 1911.
247 RGS, Correspondence Block 1911–1920. John S. Keltie to Semple, 22 April 1911.
248 Semple: *American History*, title page.
249 Fitzpatrick, Ellen F., *History's Memory: Writing America's Past, 1880–1980* (Cambridge, MA: Harvard University Press, 2004), p. 77.

CHAPTER 3

1   Hardwick, Lorna, *Reception Studies* (Oxford: Oxford University Press, 2003), p. 4.
2   See, for example, Brush, Stephen G., 'The reception of Mendeleev's periodic law in America and Britain', *Isis* 87/4 (1996), pp. 595–628; Conlin, Michael F., 'The popular and scientific reception of the Foucault Pendulum in the United States', *Isis* 90/2 (1999), pp. 181–204; and Russell, Colin A., 'The reception of Newtonianism in Europe', in D. Goodman and C. A. Russell (eds), *The Rise of Scientific Europe 1500–1800* (Sevenoaks: Hodder & Stoughton, 1991), pp. 253–78.
3   See, for example, Fyfe, Aileen, 'The reception of William Paley's *Natural Theology* in the University of Cambridge', *British Journal for the History of Science* 30/3 (1997), pp. 321–35; Fyfe, Aileen, *Science and Salvation: Evangelical Popular Science Publishing in Victorian Britain* (Chicago and London: University of Chicago Press, 2004); and Livingstone, David N., 'Science, religion and the geography of reading: Sir William Whitla and the editorial staging of Isaac Newton's writings on biblical prophecy', *The British Journal for the History of Science* 36/1 (2003), pp. 27–42.
4   Livingstone, David N., 'Text, talk and testimony: geographical reflections on scientific habits. An afterword', *British Journal for the History of Science* 38/1 (2005), pp. 93–100, p. 99.
5   Rupke, Nicolaas A., 'A geography of enlightenment: the critical reception of Alexander von Humboldt's Mexico work', in D. N. Livingstone and C. W. J. Withers (eds), *Geography and Enlightenment* (Chicago and London: University of Chicago Press, 1999), pp. 319–43, p. 324.
6   For a discussion of the epistemic difficulties associated with the use of published reviews in this context, see Maddrell, Avril M. C., *Complex Locations: Women's Geographical Work in the UK 1850–1970* (Chichester: Wiley-Blackwell, 2009), pp. 23–24.
7   Semple, Ellen C., *Influences of Geographic Environment on the Basis of Ratzel's System of Anthropo-geography* (New York: Henry Holt and Company, 1911), p. v.
8   Semple: *Influences of Geographic Environment*, p. v.
9   Semple: *Influences of Geographic Environment*, p. v.
10 Semple: *Influences of Geographic Environment*, p. vi.
11 Semple: *Influences of Geographic Environment*, p. v.

12 Wright, John K., 'Miss Semple's "Influences of geographic environment" notes towards a bibliobiography', *Geographical Review* 52/3 (1962), pp. 346–61, p. 347.
13 Semple: *Influences of Geographic Environment*, p. vi.
14 Semple: *Influences of Geographic Environment*, p. vii.
15 Semple: *Influences of Geographic Environment*, p. vii.
16 Semple: *Influences of Geographic Environment*, p. vii.
17 Semple: *Influences of Geographic Environment*, p. vii.
18 Semple: *Influences of Geographic Environment*, p. 1.
19 Semple: *Influences of Geographic Environment*, p. 2.
20 Semple: *Influences of Geographic Environment*, p. 11.
21 Semple: *Influences of Geographic Environment*, p. 11.
22 Semple: *Influences of Geographic Environment*, p. 35. Italicization in original.
23 Semple: *Influences of Geographic Environment*, pp. 41–43. Italicization in original.
24 Wright, John K., *Human Nature in Geography* (Cambridge, MA: Harvard University Press, 1966), p. 191.
25 *The Post-Standard*, 11 March 1911.
26 UK, 46M139, Box 10. Scrapbook, 1895–1932.
27 Princeton University, Department of Rare Books and Special Collections (hereafter PU), C0100, Box 115, Folder 12. Semple to Edward N. Bristol, 4 January 1912.
28 *The Publishers Weekly*, 17 June 1911.
29 *The Publishers Weekly*, 17 June 1911.
30 Nelson, Jack, '*New Orleans Times-Picayune* series on racism', in T. Rosensteil and A. S. Mitchell (eds), *Thinking Clearly: Cases in Journalistic Decision-Making* (New York: Columbia University Press, 2003), p. 212.
31 Dabney, Thomas E., *One Hundred Great Years: The Story of the* Times-Picayune *From its Founding to 1940* (Baton Rouge: Louisiana State University Press, 1944), p. 378.
32 *Daily-Picayune*, 18 June 1911.
33 *Daily-Picayune*, 18 June 1911.
34 *The Sun*, 24 June 1911.
35 *The Sun*, 24 June 1911.
36 *The Sun*, 24 June 1911.
37 *The Sun*, 24 June 1911.
38 *The Sun*, 24 June 1911.
39 *The Sun*, 24 June 1911.
40 *The Sun*, 24 June 1911.
41 *The Sun*, 24 June 1911.
42 *The Sun*, 24 June 1911.
43 *The Sun*, 24 June 1911.
44 *Boston Evening Transcript*, 5 July 1911.
45 Chorley, Richard J., Dunn, Antony J., and Beckinsale, Robert P., *The History of the Study of Landforms or the Development of Geomorphology, Volume 2: The Life and Work of William Morris Davis* (London: Methuen, 1973).
46 *Boston Evening Transcript*, 5 July 1911.
47 *Boston Evening Transcript*, 5 July 1911.
48 *Boston Evening Transcript*, 5 July 1911.
49 *Boston Evening Transcript*, 5 July 1911.
50 *Boston Evening Transcript*, 5 July 1911.
51 Chamberlin, Joseph E., The Boston Transcript: *A History of its First Hundred Years* (Boston: Houghton Mifflin, 1930).

52 Butcher, Philip (ed.), *The William Stanley Braithwaite Reader* (Ann Arbor: University of Michigan Press, 1972); and Szefel, Lisa, 'Encouraging verse: William S. Braithwaite and the poetics of race', *The New England Quarterly* 74/1 (2001), pp. 32–61.

53 Negri, Paul (ed.), *Great Short Poems* (Mineola, NY: Dover, 2000), p. 51.

54 Koelsch, William A., 'Seedbed of reform: Arnold Guyot and school geography in Massachusetts, 1849–1855', *The Journal of Geography* 107/2 (2008), pp. 35–42, p. 37.

55 Kroll, Barry M., 'Writing for readers: three perspectives on audience', *College Composition and Communication* 35/2 (1984), pp. 172–85, p. 173.

56 Vandenberg, Peter, 'Coming to terms: audience', *The English Journal* 84/4 (1995), pp. 79–80, p. 79.

57 Burton, David H., 'Theodore Roosevelt's Social Darwinism and views on imperialism', *Journal of the History of Ideas* 26/1 (1965), pp. 103–18; and Roosevelt, Theodore, *The Winning of the West*, 4 vols. (New York: G. P. Putnam's, 1889–1896).

58 *The Outlook*, 15 July 1911.

59 *The Outlook*, 15 July 1911.

60 Chamberlain, Houston S., *Die Grundlagen des neunzehnten Jahrhunderts*, 2 vols. (Munich: F. Bruckmann, 1899); and Chamberlain, Houston S., *The Foundations of the Nineteenth Century*, tr. J. Lees, 2 vols. (New York: John Lane, 1911).

61 Roosevelt, Theodore, *History as Literature and Other Essays* (New York: Charles Scribner's Sons, 1913), pp. 235–36.

62 Semple: *Influences of Geographic Environment*, p. vii.

63 *Providence Daily Journal*, 20 August 1911.

64 *Providence Daily Journal*, 20 August 1911.

65 *Providence Daily Journal*, 20 August 1911.

66 *Providence Daily Journal*, 20 August 1911.

67 *American Library Association Booklist*, September 1911.

68 *Boston Herald*, 2 September 1911.

69 *Boston Herald*, 2 September 1911.

70 *Boston Herald*, 2 September 1911.

71 Hooker, Richard, *The Story of an Independent Newspaper* (New York: The Macmillian Company, 1924).

72 *Springfield Daily Republican*, 7 September 1911.

73 *Springfield Daily Republican*, 7 September 1911.

74 *Springfield Daily Republican*, 7 September 1911.

75 *Springfield Daily Republican*, 7 September 1911.

76 *Springfield Daily Republican*, 7 September 1911.

77 Semple: *Influences of Geographic Environment*, p. 608.

78 Semple: *Influences of Geographic Environment*, p. 608.

79 Semple: *Influences of Geographic Environment*, p. 608.

80 *Springfield Daily Republican*, 7 September 1911.

81 For a discussion of Morse's contribution see, for example, Livingstone, David N., '"Risen into empire": moral geographies of the American Republic', in D. N. Livingstone and C. W. J. Withers (eds), *Geography and Revolution* (Chicago: University of Chicago Press, 2005); and Smith, Ben A. and Vining, James W., *American Geographers, 1784–1812: A Bio-Bibliographical Guide* (Westport, CT: Praeger Publishers, 2003); and Withers, Charles W. J., *Placing the Enlightenment: Thinking Geographically About the Age of Reason* (Chicago: University of Chicago Press, 2007).

82 *Providence Daily Journal*, 20 August 1911; *The Continent*, 24 August 1911.

83 *The Hartford Courant*, 23 September 1911.

84 Buttimer, Anne, *Society and Milieu in the French Geographic Tradition* (Chicago: Rand McNally, 1971), p. 29.

85 Berman, Jessica S., *Modernist Fiction, Cosmopolitanism, and the Politics of Community* (Cambridge: Cambridge University Press, 2001), p. 164.

86 Bronson, Judith C., 'Ellen Semple: contributions to the history of American geography' (Ph.D. diss., Saint Louis University, 1973), p. 98.

87 *Review of Reviews*, July 1911.

88 *The Outlook*, 15 July 1911.

89 RGS, Correspondence Block 1911–1920. Semple to John S. Keltie, 2 April 1911.

90 PU, C0100, Box 115, Folder 12. Treasurer to Semple, 16 November 1912.

91 PU, C0100, Box 115, Folder 12. Treasurer to Semple, 16 November 1912.

92 PU, C0100, Box 115, Folder 12. Treasurer to Semple, 16 November 1912.

93 *The Bookseller*, 29 September 1911.

94 *The Bookseller*, 29 September 1911.

95 *The Bookseller*, 29 September 1911.

96 *The Morning Post*, 5 October 1911.

97 Haller, John S., 'The species problem: nineteenth-century concepts of racial inferiority in the origin of man controversy', *American Anthropologist* 72/6 (1970), pp. 1319–29; Livingstone, David N., *Adam's Ancestors: Race, Religion, and the Politics of Human Origins* (Baltimore: Johns Hopkins University Press, 2008); and Livingstone, David N., 'Geographical inquiry, rational religion, and moral philosophy: enlightenment discourses on the human condition', in D. N. Livingstone and C. W. J. Withers (eds), *Geography and Enlightenment* (Chicago and London: University of Chicago Press, 1999), pp. 93–119.

98 *The Morning Post*, 5 October 1911.

99 *The Morning Post*, 5 October 1911.

100 *The Morning Post*, 5 October 1911.

101 *The Morning Post*, 5 October 1911.

102 Hindle, Wilfrid H., The Morning Post, *1772–1937: Portrait of a Newspaper* (London: G. Routledge & Sons, 1937).

103 Ben-Itto, Hadassa, *The Lie That Wouldn't Die:* The Protocols of the Elders of Zion (London: Valentine Mitchell, 2005).

104 Wilson, Keith M., 'The "Protocols of Zion" and the "Morning Post", 1919–1920', *Patterns of Prejudice* 9/3 (1985), pp. 5–14; and Wilson, Keith M., *A Study in the History and Politics of* The Morning Post, *1905–1926* (Lewiston: E. Mellen Press, 1990).

105 Wheatley, Michael, *Nationalism and the Irish Party: Provincial Ireland 1910–1916* (Oxford: Oxford University Press, 2005).

106 *Irish Times*, 20 October 1911.

107 *Irish Times*, 20 October 1911.

108 *Irish Times*, 20 October 1911.

109 *Irish Times*, 20 October 1911.

110 *Irish Times*, 20 October 1911.

111 Semple: *Influences of Geographic Environment*, p. 298.

112 Semple: *Influences of Geographic Environment*, p. 435.

113 Semple: *Influences of Geographic Environment*, p. 435.

114 *Irish Times*, 20 October 1911.

115 *Irish Times*, 20 October 1911.

116 *Influences of Geographic Environment*, American Geographical Society Library, GF31 .S5.

117 American Geographical Society Archives (hereafter AGSA). Isaiah Bowman Papers, Folder 'Semple, Ellen Churchill 1920–32'. Semple to John K. Wright, 1 February [1922].

118 PU, C0100, Box 115, Folder 12. Semple to Edward N. Bristol, 4 January 1912.

119 Millett, Frederick B., *Contemporary American Authors: A Critical Survey and 219 Bio-Bibliographies* (London: Harrap, 1940).

120 The Newberry Library (hereafter NL), Midwest MS Dell, 1908–1969, Box 11, Folder 426. Semple to Dell, [1911].

121 Dumenil, Lynn, 'Love, literature, and politics in the machine age. Review of *Floyd Dell: The Life and Times of an American Rebel*, by Douglas Clayton', *Reviews in American History* 23/4 (1995), pp. 699–703, p. 700.

122 NL, Midwest MS Dell, 1908–1969, Box 11, Folder 426. Semple to Dell, 18 November 1911.

123 Weber, Ronald, *The Midwestern Ascendancy in American Writing* (Bloomington: Indiana University Press, 1992), p. 82.

124 *Chicago Evening Post*, 10 November 1911.

125 *Chicago Evening Post*, 10 November 1911.

126 *Chicago Evening Post*, 10 November 1911.

127 *Chicago Evening Post*, 10 November 1911.

128 Kohlstedt, Sally G., 'Concepts of place in museum space at the Smithsonian Institution's National Museum' (paper presented at the Geographies of Nineteenth-Century Science, University of Edinburgh, 2007); and Kohlstedt, Sally G., '"Thoughts in things": modernity, history, and North American museums', *Isis* 96/4 (2005), pp. 586–601.

129 Mason, Otis T., 'Resemblances in arts widely separated', *The American Naturalist* 20/3 (1886), pp. 246–51, p. 248.

130 Buettner-Janusch, John, 'Boas and Mason: particularism versus generalization', *American Anthropologist* 59/2 (1957), pp. 318–24.

131 *Chicago Evening Post*, 10 November 1911.

132 *Chicago Evening Post*, 10 November 1911.

133 *Chicago Evening Post*, 10 November 1911.

134 *Chicago Evening Post*, 10 November 1911.

135 Golding, Alan C., '*The Dial*, *The Little Review*, and the dialogics of modernism', *American Periodicals* 15/1 (2005), pp. 42–55, p. 44.

136 Kirby, Harold, 'Charles Atwood Kofoid 1865–1947', *Science* 106/2759 (1947), pp. 462–63, p. 462.

137 Kirby: 'Charles Atwood Kofoid', p. 462.

138 Kofoid, Charles A., 'Environment as a force in human action', *The Dial* 51/610 (1911), pp. 398–99, p. 399.

139 Kofoid: 'Environment as a force in human action', p. 399.

140 Kofoid: 'Environment as a force in human action', p. 398.

141 Kofoid: 'Environment as a force in human action', p. 399.

142 Kofoid: 'Environment as a force in human action', p. 399.

143 Kofoid: 'Environment as a force in human action', p. 399.

144 Semple: *Influences of Geographic Environment*, p. 2.

145 Semple: *Influences of Geographic Environment*, p. 2.

146 Kofoid: 'Environment as a force in human action', p. 399.

147 *The Nation*, 21 December 1911.

148 *The Nation*, 21 December 1911.

149 Lloyd, Genevieve, *The Man of Reason: 'Male' and 'Female' in Western Philosophy* (London: Methuen, 1984).

150 *The Nation*, 21 December 1911.

151 *The Nation*, 21 December 1911.

152 *The Nation*, 21 December 1911.

153 *The Nation*, 21 December 1911.

154 *The Nation*, 21 December 1911.

155 *The Nation*, 21 December 1911.

156 *The Nation*, 21 December 1911.

157 *The Nation*, 21 December 1911.

158 PU, C0100, Box 115, Folder 12. Semple to Edward N. Bristol, 12 February 1912.

159 PU, C0100, Box 115, Folder 12. Edward N. Bristol to Semple, 11 March 1912.

160 *The Sun*, 24 June 1911.

161 *Newark News*, 23 December 1911.

162 *The Nation*, 21 December 1911.

163 *The Saturday Review*, 30 December 1911.

164 *The Manchester Guardian*, 29 January 1912.

165 *The Manchester Guardian*, 29 January 1912.

166 'Review of *Influences of Geographic Environment*, by Ellen C. Semple', *The Journal of Geography* 10/1 (1911), pp. 33–34.

167 Davis, William M. and Dodge, Richard E., 'The *Journal of School Geography*', *Science* 5/118 (1897), pp. 551–53.

168 See, for example, Carney, Frank, 'Geographic influences in the development of Ohio', *The Journal of Geography* 9/7 (1911), pp. 169–74; Chamberlain, James F., 'Geographic influences in the development of California', *The Journal of Geography* 9/10 (1911), pp. 253–61; Dodge, Richard E., 'Man and his geographic environment', *The Journal of Geography* 8/8 (1910), pp. 179–87; Dryer, Charles R., 'Geographic influences in the development of India', *The Journal of Geography* 9/1 (1910), pp. 17–22; Goode, J. Paul, 'The human response to the physical environment', *The Journal of Geography* 3/7 (1904), pp. 333–43; Whitbeck, Ray H., 'Geographic influences in the development of New York State', *The Journal of Geography* 9/5 (1911), pp. 119–24; Whitbeck, Ray H., 'Mountains in their influence on man and his activities', *The Journal of Geography* 9/2 (1910), pp. 54–55; and Whitbeck, Ray H., 'Response to surroundings – a geographic principle', *The Journal of Geography* 3/9 (1904), pp. 409–12.

169 'Review of *Influences*', *The Journal of Geography*, p. 33.

170 'Review of *Influences*', *The Journal of Geography*, p. 33.

171 'Review of *Influences*', *The Journal of Geography*, p. 33.

172 'Review of *Influences*', *The Journal of Geography*, p. 33.

173 'Review of *Influences*', *The Journal of Geography*, p. 33.

174 'Review of *Influences*', *The Journal of Geography*, p. 33.

175 'Review of *Influences*', *The Journal of Geography*, p. 33.

176 'Review of *Influences*', *The Journal of Geography*, p. 33.

177 Semple: *Influences of Geographic Environment*, p. vii.

178 'Review of *Influences*', *The Journal of Geography*, p. 34.

179 'Human geography. Review of *Influences of Geographic Environment*, by Ellen C. Semple', *Nature* 88/2195 (1911), p. 101.

180 'Human geography', *Nature*, p. 101.

181 'Human geography', *Nature*, p. 101.

182 'Human geography', *Nature*, p. 101.

183 'Human geography', *Nature*, p. 101.

184 'Human geography', *Nature*, p. 101.

185 'Human geography', *Nature*, p. 101.

186 Whitbeck, Ray H., 'Review of *Influences of Geographic Environment*, by Ellen C. Semple', *Bulletin of the American Geographical Society* 43/12 (1911), pp. 937–39.

187 Trewartha, Glenn T., 'Geography at Wisconsin', *Annals of the Association of American Geographers* 69/1 (1979), pp. 16–21, p. 18.
188 Brigham, Albert P., 'Memoir of Ralph Stockman Tarr', *Annals of the Association of American Geographers* 3 (1913), pp. 93–98; Engeln, Oscar D. von, 'Ralph Stockman Tarr', *Bulletin of the American Geographical Society* 44/4 (1912), pp. 283–85; and Tarr, Ralph S., 'Review of *American History and its Geographic Conditions*, by Ellen C. Semple', *Bulletin of the American Geographical Society* 35/5 (1903), pp. 566–70.
189 Williams, Frank E., 'Ray Hughes Whitbeck: geographer, teacher, and man', *Annals of the Association of American Geographers* 30/3 (1940), pp. 210–18, p. 212.
190 Whitbeck: 'Review of *Influences of Geographic Environment*', p. 937.
191 Whitbeck: 'Review of *Influences of Geographic Environment*', p. 937.
192 Whitbeck: 'Review of *Influences of Geographic Environment*', p. 938.
193 Whitbeck: 'Review of *Influences of Geographic Environment*', p. 938.
194 Whitbeck: 'Review of *Influences of Geographic Environment*', p. 938.
195 Whitbeck: 'Review of *Influences of Geographic Environment*', p. 938.
196 Whitbeck: 'Review of *Influences of Geographic Environment*', p. 938.
197 Mayhew, Robert J., 'Mapping science's imagined community: geography as a Republic of Letters, 1600–1800', *British Journal for the History of Science* 38/1 (2005), pp. 73–92, p. 77.
198 Whitbeck: 'Review of *Influences of Geographic Environment*', p. 939.
199 Libby, Orin G., 'Review of *Influences of Geographic Environment*, by Ellen C. Semple', *The American Historical Review* 17/2 (1912), pp. 355–57.
200 Jacobs, Wilbur R., 'Colonial origins of the United States: the Turnerian view', *The Pacific Historical Review* 40/1 (1971), pp. 21–37.
201 Libby: 'Review of *Influences of Geographic Environment*', p. 355.
202 Libby: 'Review of *Influences of Geographic Environment*', p. 355.
203 Libby: 'Review of *Influences of Geographic Environment*', p. 355.
204 Libby: 'Review of *Influences of Geographic Environment*', p. 356.
205 Libby: 'Review of *Influences of Geographic Environment*', p. 356.
206 Libby: 'Review of *Influences of Geographic Environment*', p. 356.
207 Libby: 'Review of *Influences of Geographic Environment*', p. 357.
208 Libby: 'Review of *Influences of Geographic Environment*', p. 357.
209 Libby: 'Review of *Influences of Geographic Environment*', p. 357.
210 'Review of *Influences*', *The Journal of Geography*, p. 33; and Libby, 'Review of *Influences of Geographic Environment*', p. 355.
211 RGS, Correspondence Block 1911–1920. John S. Keltie to Semple, 22 April 1911.
212 Barnes, Trevor J., 'Performing economic geography: two men, two books, and a cast of thousands', *Environment and Planning A* 34/3 (2002), pp. 487–512; and Withers, Charles W. J., *Geography, Science and National Identity: Scotland Since 1520* (Cambridge: Cambridge University Press, 2001).
213 RGS, Correspondence Block 1881–1911. Clements Markham to John S. Keltie, 10 September 1911.
214 Close, Charles F., 'The position of geography', *The Geographical Journal* 38/4 (1911), pp. 404–13.
215 Close: 'Position of geography', p. 409.
216 Withers, Charles W. J., Finnegan, Diarmid A., and Higgitt, Rebekah, 'Geography's other histories? Geography and science in the British Association for the Advancement of Science, 1831–c.1933', *Transactions of the Institute of British Geographers* 31/4 (2006), pp. 433–51.

217 Genthe, Martha K., 'Comment on Colonel Close's address on the purpose and position of geography', *Bulletin of the American Geographical Society* 44/1 (1912), pp. 27–38.

218 RGS, Correspondence Block 1881–1911. George G. Chisholm to John S. Keltie, 18 September 1911

219 RGS, Correspondence Block 1881–1911. Hugh Robert Mill to John S. Keltie, 23 September 1911.

220 MacLean, Kenneth, 'George Goudie Chisholm 1850–1930', in T. W. Freeman (ed.), *Geographers: Biobibliographical Studies*, vol. 12 (London: Mansell, 1988), pp. 21–33; and Wise, Michael J., 'A university teacher of geography', *Transactions of the Institute of British Geographers* 66 (1975), pp. 1–16.

221 Chisholm, George G., 'Some recent contributions to geography', *Scottish Geographical Magazine* 27 (1911), pp. 561–73.

222 Chisholm, George G., 'The meaning and scope of geography', *Scottish Geographical Magazine* 24 (1908), pp. 561–75, p. 575.

223 Chisholm, George G., 'Miss Semple on the influences of geographical environment. Review of *Influences of Geographic Environment* by Ellen C. Semple', *The Geographical Journal* 39/1 (1912), pp. 31–37, p. 31.

224 Chisholm: 'Miss Semple', p. 31.

225 Chisholm: 'Miss Semple', p. 31.

226 Chisholm: 'Miss Semple', p. 31.

227 Chisholm: 'Miss Semple', p. 32.

228 Chisholm: 'Miss Semple', p. 32.

229 Chisholm: 'Miss Semple', p. 32.

230 Chisholm: 'Miss Semple', p. 32.

231 Chisholm: 'Miss Semple', p. 34.

232 Chisholm: 'Miss Semple', p. 32.

233 Chisholm: 'Miss Semple', p. 33.

234 'Review of *Influences of Geographic Environment*, by Ellen C. Semple', *Scottish Geographical Magazine* 27 (1912), pp. 160–62, p. 160.

235 Bell, Morag, 'Reshaping boundaries: international ethics and environmental consciousness in the early twentieth century', *Transactions of the Institute of British Geographers* 23/2 (1998), pp. 151–75; Freeman, Thomas W., 'The Manchester and Royal Scottish Geographical Societies', *The Geographical Journal* 150/1 (1984), pp. 55–62; and Maddrell, Avril M. C., 'Scientific discourse and the geographical work of Marion Newbigin', *Scottish Geographical Magazine* 113/1 (1997), pp. 33–41.

236 Bowen, Emrys G., Carter, Harold, and Taylor, James A. (eds), *Geography at Aberystwyth: Essays Written on the Occasion of the Departmental Jubilee 1917–18 – 1967–68* (Cardiff: University of Wales Press, 1968); Freeman, Thomas W., 'Herbert John Fleure 1877–1969', in T. W. Freeman (ed.), *Geographers: Biobibliographical Studies*, vol. 11 (London: Mansell, 1987), pp. 35–51; and Garnett, Alice, 'Herbert John Fleure: 1877–1969', *Biographical Memoirs of Fellows of the Royal Society* 16 (1970), pp. 253–78.

237 National Library of Wales (hereafter NLW), A1982/3 1977152, item 7.

238 Freeman: 'Herbert John Fleure', p. 38.

239 AGSL. Herbert J. Fleure to John K. Wright, 31 May 1961.

240 AGSL. Herbert J. Fleure to John K. Wright, 31 May 1961.

241 Fleure, Herbert J., 'Review of *Influences of Geographic Environment*, by Ellen C. Semple', *The Geographical Teacher* 7/35 (1913), pp. 65–68.

242 Fleure: 'Review of *Influences of Geographic Environment*", p. 68.

243 Roorbach, George B., 'Review of *Influences of Geographic Environment*, by Ellen C. Semple', *Annals of the American Academy of Political and Social Science* 41 (1912), pp. 350–51.
244 Roorbach: 'Review of *Influences of Geographic Environment*', p. 350.
245 Roorbach: 'Review of *Influences of Geographic Environment*', p. 350.
246 'George Byron Roorbach', *Geographical Review* 24/3 (1934), p. 506.
247 Roorbach: 'Review of *Influences of Geographic Environment*', p. 350.
248 Columbia University, Rare Book and Manuscript Library (hereafter CoU), Ms Coll Robinson, E. V., Box 3, Binder 5, Section 10. Davis R. Dewey to Robinson 15 December 1911.
249 PU, C0100, Box 115, Folder 12. Semple to Joseph Vogelius, 26 May 1911.
250 Fellmann, Jerome D., 'Myth and reality in the origin of American economic geography', *Annals of the Association of American Geographers* 76/3 (1986), pp. 313–30, p. 327.
251 Dodge, Richard E., 'Edward Van Dyke Robinson', *Annals of the Association of American Geographers* 6 (1916), p. 120.
252 Robinson, Edward van Dyke, 'Review of *Influences of Geographic Environment*, by Ellen C. Semple', *The American Economic Review* 2/2 (1912), pp. 338–40.
253 Robinson: 'Review of *Influences of Geographic Environment*', p. 338. My italicization.
254 Robinson: 'Review of *Influences of Geographic Environment*', p. 339.
255 Robinson: 'Review of *Influences of Geographic Environment*', p. 339.
256 Robinson: 'Review of *Influences of Geographic Environment*', p. 339.
257 Robinson: 'Review of *Influences of Geographic Environment*', p. 339.
258 Robinson: 'Review of *Influences of Geographic Environment*', p. 340.
259 Taft, Donald R., 'Alvan A. Tenney: a tribute', *American Sociological Review* 2/1 (1937), pp. 67–68.
260 Bannister, Robert C., *Sociology and Scientism: The American Quest for Objectivity, 1880–1940* (Chapel Hill: University of North Carolina Press, 1987), p. 77.
261 Bannister: *Sociology and Scientism*, p. 78.
262 Semple, *Influences of Geographic Environment*, p. vii; and Tenney, Alvan A., 'Review of *Influences of Geographic Environment*, by Ellen C. Semple', *Political Science Quarterly* 27/2 (1912), pp. 345–48, p. 346.
263 Tenney: 'Review of *Influences of Geographic Environment*', p. 346.
264 Tenney: 'Review of *Influences of Geographic Environment*', p. 346.
265 Tenney: 'Review of *Influences of Geographic Environment*', p. 346.
266 Tenney: 'Review of *Influences of Geographic Environment*', p. 347.
267 Tenney: 'Review of *Influences of Geographic Environment*', p. 347.
268 Tenney: 'Review of *Influences of Geographic Environment*', p. 348.
269 Tenney: 'Review of *Influences of Geographic Environment*', p. 348.
270 Semple: *Influences of Geographic Environment*, p. v.
271 Kish, George, 'Roberto Almagià', *Geographical Review* 52/4 (1962), pp. 611–12.
272 'Giuseppe Dalla Vedova', *The Geographical Journal* 55/2 (1920), pp. 157–58, p. 158; and Luzzana Caraci, Ilaria, 'Modern geography in Italy: from the archives to environmental management', in G. S. Dunbar (ed.), *Geography: Discipline, Profession and Subject Since 1870* (Dordrecht: Kluwer, 2001), pp. 121–51.
273 Almagià, Roberto, 'Review of *Influences of Geographic Environment*, by Ellen C. Semple', *Bollettino della Società Geographica Italiana* 5/1 (1912), pp. 550–54, p. 551. Translated from original.
274 Almagià: 'Review of *Influences of Geographic Environment*', p. 551.
275 Almagià: 'Review of *Influences of Geographic Environment*', p. 551.
276 Almagià: 'Review of *Influences of Geographic Environment*', p. 551.
277 Almagià: 'Review of *Influences of Geographic Environment*', p. 552.

278 Almagià: 'Review of *Influences of Geographic Environment*', p. 553.
279 Almagià: 'Review of *Influences of Geographic Environment*', p. 553.
280 Almagià: 'Review of *Influences of Geographic Environment*', pp. 553–54.
281 Almagià: 'Review of *Influences of Geographic Environment*', p. 554.
282 Almagià: 'Review of *Influences of Geographic Environment*', p. 554.
283 *Boston Herald*, 2 September 1911; and Schlüter, Otto, 'Review of *Influences of Geographic Environment*, by Ellen C. Semple', *Petermanns Geographische Mitteilungen* 58 (1912), p. 166.
284 Ritter, Carl, *Comparative Geography*, tr. W. L. Gage (Philadelphia: J. B. Lippincott & Co., 1865).
285 Schlüter: 'Review of *Influences of Geographic Environment*', p. 166. Translated from original.
286 Schlüter: 'Review of *Influences of Geographic Environment*', p. 166.
287 Schlüter: 'Review of *Influences of Geographic Environment*', p. 166.
288 Schlüter: 'Review of *Influences of Geographic Environment*', p. 166.
289 Keller, Albert G., 'Review of *Influences of Geographic Environment*, by Ellen C. Semple', *The Yale Review* 1 (1912), pp. 331–34, p. 333.
290 Keller: 'Review of *Influences of Geographic Environment*', p. 333.
291 Livingstone, David N., 'Science, site and speech: scientific knowledge and the spaces of rhetoric', *History of the Human Sciences* 20/2 (2007), pp. 71–98, p. 71.

CHAPTER 4

1 PU, C0100, Box 115, Folder 12. Semple to Edward N. Bristol, 13 March 1911.
2 RGS, Correspondence Block 1911–1920. Semple to John S. Keltie, 11 November 1911.
3 RGS, Correspondence Block 1911–1920. Semple to John S. Keltie, 2 April 1911.
4 Secord, James A., 'How scientific conversation became shop talk', *Transactions of the Royal Historical Society* 17 (2007), pp. 129–56, p. 129.
5 Secord: 'How scientific conversation became shop talk', p. 129.
6 See, for example, Lightman, Bernard, 'Lecturing in the spatial economy of science', in A. Fyfe and B. Lightman (eds), *Science in the Marketplace: Nineteenth-Century Sites and Experiences* (Chicago and London: University of Chicago Press, 2007); Livingstone, David N., 'Text, talk and testimony: geographical reflections on scientific habits. An afterword', *British Journal for the History of Science* 38/1 (2005), pp. 93–100; Scott, Donald M., 'The popular lecture and the creation of a public in mid-nineteenth-century America', *The Journal of American History* 66/4 (1980), pp. 791–809; and van Wyhe, John, 'The diffusion of phrenology through public lecturing', in A. Fyfe and B. Lightman (eds), *Science in the Marketplace: Nineteenth-Century Sites and Experiences* (Chicago and London: University of Chicago Press, 2007), pp. 60–96.
7 Finnegan, Diarmid A., 'Natural history societies in late Victorian Scotland and the pursuit of local civic science', *British Journal for the History of Science* 38/1 (2005), pp. 53–72.
8 Finnegan, Diarmid A., '"A free and liberal tone": scientific speech and intellectual culture in mid-Victorian Edinburgh' (paper presented at the Geographies of Nineteenth-Century Science, University of Edinburgh, 2007).
9 Livingstone, David N., 'Science, site and speech: scientific knowledge and the spaces of rhetoric', *History of the Human Sciences* 20/2 (2007), pp. 71–98, p. 71.
10 Lyell, Charles, *Travels in North America; With Geological Observations on the United States, Canada, and Nova Scotia*, vol. 1 (London: John Murray, 1845), pp. 110–11.

11 RGS, Correspondence Block 1911–1920. Semple to John S. Keltie, 30 October 1912.
12 *The Evening Post*, 9 November 1912.
13 Bell, Morag and McEwan, Cheryl, 'The admission of women fellows to the Royal Geographical Society, 1892–1914; the controversy and the outcome', *The Geographical Journal* 162/3 (1996), pp. 295–312; and Domosh, Mona, 'Toward a feminist historiography of geography', *Transactions of the Institute of British Geographers* 16/1 (1991), pp. 95–104.
14 Bird, Isabella L., *Unbeaten Tracks in Japan: An Account of Travels in the Interior, Including Visits to the Aborigines of Yezo and the Shrines of Nikkô and Isé* (London: John Murray, 1880).
15 See, for example, Semple, Ellen C., 'Japanese colonial methods', *Bulletin of the American Geographical Society* 45/4 (1913), pp. 255–75.
16 Domosh, Mona, 'Beyond the frontiers of geographical knowledge', *Transactions of the Institute of British Geographers* 16/4 (1991), pp. 488–89.
17 RGS, Correspondence Block 1911–1920. Semple to John S. Keltie, 19 August 1911.
18 Birkett, Dea, *Spinsters Abroad: Victorian Lady Explorers* (Oxford: Basil Blackwell, 1989), p. 57.
19 RGS, Correspondence Block 1911–1920. Semple to John S. Keltie, 19 August 1911.
20 *The Evening Post*, 9 November 1912.
21 *Springfield Daily Republican*, 7 September 1911.
22 *The Evening Post*, 9 November 1912.
23 PU, C0100, Box 115, Folder 12. Semple to Edward N. Bristol, 4 January 1912.
24 *The Evening Post*, 9 November 1912.
25 RGS, Correspondence Block 1911–1920. Semple to John S. Keltie, 23 January 1912.
26 RGS, Correspondence Block 1911–1920. John S. Keltie to Semple, 6 September 1911.
27 RGS, Correspondence Block 1911–1920. Semple to John S. Keltie, 11 November 1911.
28 RGS, Correspondence Block 1911–1920. Semple to John S. Keltie, 11 November 1911.
29 RGS, Correspondence Block 1911–1920. Semple to John S. Keltie, 11 November 1911.
30 RGS, Correspondence Block 1911–1920. John S. Keltie to Semple, 11 December 1911.
31 Bladen, Wilford A. and Karan, Pradyumna P. (eds), *The Evolution of Geographic Thought in America: A Kentucky Root* (Dubuque, IA: Kendall/Hunt, 1983).
32 *The Evening Post*, 9 November 1912.
33 Waller, Philip, *Writers, Readers, and Reputations: Literary Life in Britain 1870–1918* (Oxford: Oxford University Press, 2006).
34 RGS, Correspondence Block 1911–1920. Semple to John S. Keltie, 5 July 1912.
35 Carter, C. C., and McGregor, C., 'Long vacation course at the School of Geography, Oxford: the work of the course', *The Geographical Teacher* 1 (1901–1902), pp. 172–74; Freeman, Thomas W., *A History of Modern British Geography* (London: Longman, 1980); and Kearns, Gerry, 'Halford John Mackinder 1861–1947', in T. W. Freeman (ed.), *Geographers: Biobibliographical Studies*, vol. 9 (London: Mansell, 1985), pp. 71–86.
36 Campbell, John A. and Livingstone, David N., 'Neo-Lamarckism and the development of geography in the United States and Great Britain', *Transactions of the Institute of British Geographers* 8/3 (1983), pp. 267–94, p. 283.

37 Dickinson, Robert E., *Regional Concept: The Anglo-American Leaders* (London: Routledge and Kegan Paul, 1976); and Williams, Frank E., 'Geographer-envoy from America to Europe', *Annals of the Association of American Geographers* 20/2 (1930), pp. 86–90.

38 Baigent, Elizabeth, 'Herbertson, Andrew John (1865–1915)', in H. C. G. Matthew and B. Harrison (eds), *Oxford Dictionary of National Biography* (Oxford: Oxford University Press, 2004).

39 Meller, Helen, 'Geddes, Sir Patrick (1854–1932)', in H. C. G. Matthew and B. Harrison (eds), *Oxford Dictionary of National Biography* (Oxford: Oxford University Press, 2004).

40 Herbertson, Andrew J., 'The major natural regions: an essay in systematic geography', *The Geographical Journal* 25/3 (1905), pp. 300–10.

41 Watson, J. Wreford, 'The sociological aspect of geography', in G. Taylor (ed.), *Geography in the Twentieth Century* (London: Methuen, 1957), p. 467.

42 Crone, Gerald R., 'British geography in the twentieth century', *The Geographical Journal* 130/2 (1964), pp. 197–220, p. 202.

43 Oxford University Archives (hereafter OUA), GE 4/1; and GE 5.

44 CU, B4-18-11. Semple to Wallace W. Atwood, 29 August 1922; and NLW, A1982/3 1977152, item 7.

45 Fleure, Herbert J., 'The later developments in Herbertson's thought: a study in the application of Darwin's ideas', *Geography* 37 (1952), pp. 97–103, p. 99.

46 For an indication of how, quite, Geddes and Fleure differed, see, for example, Fleure, 'Developments in Herbertson's thought'; Fleure, Herbert J., 'Recollections of A. J. Herbertson', *Geography* 50 (1965), pp. 348–49; Freeman, Thomas W., 'Herbert John Fleure 1877–1969', in T. W. Freeman (ed.), *Geographers: Biobibliographical Studies*, vol. 11 (London: Mansell, 1987), pp. 35–51; Garnett, Alice, 'Herbert John Fleure: 1877–1969', *Biographical Memoirs of Fellows of the Royal Society* 16 (1970), pp. 253–78; Geddes, Patrick, 'The influence of geographical conditions on social development', *The Geographical Journal* 12/6 (1898), pp. 580–86; and Renwick, Chris, 'The practice of Spencerian science: Patrick Geddes's biosocial program, 1876–1889', *Isis* 100/1 (2009), pp. 36–57.

47 Graham, Brian J., 'The search for the common ground: Estyn Evans's Ireland', *Transactions of the Institute of British Geographers* 19/2 (1994), pp. 183–201, p. 187.

48 HUA, HUG 4877.412. Semple to Whittlesey, 4 April 1929.

49 AGSL, Herbert J. Fleure to John K. Wright, 31 May 1961.

50 AGSL, Herbert J. Fleure to John K. Wright, 3 June 1961.

51 Fleure, Herbert J. and James, T. C., 'Geographical distribution of anthropological types in Wales', *The Journal of the Royal Anthropological Institute of Great Britain and Ireland* 46 (1916), pp. 35–153, p. 36.

52 RGS, Correspondence Block 1911–1920. Semple to John S. Keltie, 5 July 1912.

53 PU, C0100, Box 115, Folder 12. Semple to Treasurer, 8 November 1912; and OUA, GE 4/1.

54 PU, C0100, Box 115, Folder 12. Semple to Treasurer, 8 November 1912.

55 *The Oxford Times*, 17 August 1912.

56 *The Evening Post*, 9 November 1912.

57 *The Oxford Times*, 17 August 1912.

58 *The Oxford Times*, 17 August 1912.

59 *The Oxford Times*, 17 August 1912.

60 Keltie, John S., *The Position of Geography in British Universities*, W. L. G. Joerg (ed.) (New York: Oxford University Press, 1921).

61 Bell and McEwan: 'Admission of women fellows', p. 303.

62 Dagg, Anne I., *The Feminine Gaze: A Canadian Compendium of Non-Fiction Women Authors and Their Books, 1836–1945* (Waterloo, ON: Wilfrid Laurier University Press, 2001).

63 Linehan, Thomas, *British Fascism, 1918–39: Parties, Ideology and Culture* (Manchester: Manchester University Press, 2000).

64 UK, 46M139, Box 10. Scrapbook, 1895–1932. Geographical Circle programme.

65 UK, 46M139, Box 10. Scrapbook, 1895–1932. Geographical Circle programme.

66 UK, 46M139, Box 10. Scrapbook, 1895–1932. Geographical Circle programme.

67 Hevly, Bruce, 'The heroic science of glacier motion', *Osiris* 11 (1996), pp. 66–86, p. 66; Kearns, Gerry, 'The imperial subject: geography and travel in the work of Mary Kingsley and Halford Mackinder', *Transactions of the Institute of British Geographers* 22/4 (1997), pp. 450–72; and Withers, Charles W. J., 'Mapping the Niger, 1798–1832: trust, testimony and "ocular demonstration" in the late Enlightenment', *Imago Mundi* 56/2 (2004), pp. 170–93.

68 'An alpine section in a woman's club', *Bulletin of the American Geographical Society* 40/9 (1908), p. 553.

69 Lukacs, John, *Destinations Past* (Columbia: University of Missouri Press, 1994), p. 212.

70 RGS, Additional Papers, Box 93/2. Cutting from *The Guardian*, 10 July 1982; Bell and McEwan: 'Admission of women fellows'; and Maddrell, Avril M. C., *Complex Locations: Women's Geographical Work in the UK 1850–1970* (Chichester: Wiley-Blackwell, 2009).

71 RGS, Additional Papers, Box 93/2. Freshfield, Douglas W., 'Memorandum on the conduct of affairs by the Council of the Royal Geographical Society', 1894.

72 *The Scotsman*, 26 November 1912.

73 *The Morning Post*, 26 November 1912.

74 *The Morning Post*, 26 November 1912.

75 *The Morning Post*, 26 November 1912.

76 Domosh: 'Beyond the frontiers of geographical knowledge', p. 488.

77 Quoted in Anderson, Monica, *Women and the Politics of Travel, 1870–1914* (Cranbury, NJ: Fairleigh Dickinson University Press, 2006), p. 23.

78 Curzon, George N., Gowland, William, Lyde, Lionel W., and Semple, Ellen C., 'Influences of geographic conditions upon Japanese agriculture – discussion', *The Geographical Journal* 40/6 (1912), pp. 603–07, p. 603.

79 Curzon *et al.*: 'Discussion', p. 603.

80 *Louisville Herald*, 5 January 1913.

81 *Daily Express*, 5 November 1912.

82 *Daily Express*, 5 November 1912.

83 Fawcett, Charles B., 'Recent losses to British geography', *Geographical Review* 37/3 (1947), pp. 505–07, p. 506.

84 *Daily Express*, 5 November 1912.

85 Curzon *et al.*: 'Discussion', p. 607.

86 *Louisville Herald*, 5 January 1913.

87 *Louisville Herald*, 5 January 1913.

88 Searle, Geoffrey R. (ed.), *Eugenics and Politics in Britain: 1900–1914* (Leyden: Noordhoff International Publishing, 1976).

89 *The Courier-Journal*, 8 January 1963.

90 Bell and McEwan: 'Admission of women fellows', p. 302.

91 *The Times*, 2 December 1912.

92 *The Scotsman*, 26 November 1912.

93 McEwan, Cheryl, 'Gender, science and physical geography in nineteenth-century Britain', *Area* 30/3 (1998), pp. 215–23.

94 McEwan: 'Gender, science and physical geography', p. 219.
95 *The Evening Post*, 8 March 1913.
96 Nelson, Robert S., 'The slide lecture, or the work of art "history" in the age of mechanical reproduction', *Critical Inquiry* 26/3 (2000), pp. 414–34, p. 415.
97 Schwartz, Joan M., '*The Geography Lesson*: photographs and the construction of imaginative geographies', *Journal of Historical Geography* 22/1 (1996), pp. 16–45, p. 20.
98 See, for example, Krauss, Rosalind, 'Photography's discursive spaces', in J. Evans and S. Hall (eds), *Visual Culture: The Reader* (London: Sage, 1989), pp. 193–209; Ryan, James R., *Picturing Empire: Photography and the Visualization of the British Empire* (London: Reaktion Books, 1997); and Tucker, Jennifer, 'Photography as witness, detective, and impostor: visual representation in Victorian science', in Bernard Lightman (ed.), *Victorian Science in Context* (Chicago and London: University of Chicago Press, 1997), pp. 378–408.
99 Nelson: 'Slide lecture', p. 422.
100 Darwin, Leonard and Lyons, Henry G., 'Notes on Moroccan geography – discussion', *The Geographical Journal* 41/3 (1913), pp. 237–39, p. 238.
101 *Louisville Herald*, 5 January 1913.
102 *The Aberdeen Free Press*, 20 November 1912.
103 *The Aberdeen Free Press*, 20 November 1912.
104 *Dundee Advertiser*, 21 November 1912.
105 *Dundee Advertiser*, 21 November 1912.
106 *The Scotsman*, 29 November 1912; *Louisville Herald*, 5 January 1913.
107 *The Scotsman*, 23 November 1912.
108 For a discussion of the importance of the provincial lecture theatre, see, for example, Finnegan: 'Scientific speech and intellectual culture'.
109 *The Scotsman*, 13 November 1912.
110 Naylor, Simon, 'The field, the museum and the lecture hall: the spaces of natural history in Victorian Cornwall', *Transactions of the Institute of British Geographers* 27/4 (2002), pp. 494–513.
111 *Chicago Evening Post*, 1 February 1913.
112 *The Louisville Times*, 4 January 1913.
113 *The Louisville Times*, 4 January 1913.
114 *The Louisville Times*, 4 January 1913.
115 *The Louisville Times*, 4 January 1913.
116 *Louisville Herald*, 5 January 1913.
117 *The Courier-Journal*, 13 June 1903.
118 *The Louisville Times*, 4 January 1913.
119 *Boston Herald*, 6 March 1913; and Wright, John K., *Geography in the Making: The American Geographical Society 1851–1951* (New York: American Geographical Society, 1952), p. 154.
120 'Meeting of the Society', *Bulletin of the American Geographical Society* 45/4 (1913), p. 285.
121 'The Cullum Geographical Medal presented to Miss Ellen Churchill Semple', *Bulletin of the American Geographical Society* 46/5 (1914), p. 364.
122 McManis, Douglas R., 'Leading ladies at the AGS', *Geographical Review* 86/2 (1996), pp. 270–77, p. 271.
123 'Cullum Geographical Medal', p. 364.
124 'Cullum Geographical Medal', p. 364.
125 *The Washington Post*, 27 March 1914.
126 *The Washington Post*, 28 March 1914.
127 'Presentation of the Cullum Geographical Medal to Hugh Robert Mill', *Geographical Review* 20/4 (1930), pp. 669–70, p. 669.

128 PU, C0100, Box 115, Folder 12. Treasurer to United States Trust Company, 10 April 1933.
129 Wellesley College Archives (hereafter WC). *Wellesley College News*, 3 December 1914; University of Colorado at Boulder (Archives) (hereafter UCoB). *University of Colorado Bulletin* 12/2 (1915); and Western Kentucky University (Museum and Library) (hereafter WKU), *Normal Heights*, April 1917.
130 Palmieri, Patricia A., 'Here was fellowship: a social portrait of academic women at Wellesley College, 1895–1920', *History of Education Quarterly* 32/2 (1983), pp. 195–214, p. 195.
131 Elder, Eleanor S., 'Women in early geology', *Journal of Geological Education* 30/5 (1982), pp. 287–93; and Shrock, Robert R., *Geology at M.I.T., 1865–1965: A History of the First Hundred Years of Geology at Massachusetts Institute of Technology* (Cambridge, MA: The MIT Press, 1982).
132 WC. *Wellesley College Course Catalog*, 1914–1915.
133 WC. *Wellesley College Course Catalog*, 1914–1915.
134 WC. *Wellesley College Course Catalog*, 1914–1915.
135 WC. *Wellesley College News*, 3 December 1914.
136 WC, 1DB/1899–1966. Report of the Department of Geology and Geography, 1914–1915.
137 WC, 1DB/1899–1966. Report of the Department of Geology and Geography, 1914–1915.
138 Visher, Stephen S., 'Rollin D. Salisbury and geography', *Annals of the Association of American Geographers* 43/1 (1953), pp. 4–11.
139 'Dissertations in geography accepted by universities in the United States for the degree of Ph.D. as of May, 1935', *Annals of the Association of American Geographers* 25/4 (1935), pp. 211–37.
140 'Geography in the summer schools', *Bulletin of the American Geographical Society* 47/5 (1915), pp. 370–72, p. 370.
141 de Souza, Anthony R., 'Talks with teachers: Elizabeth Eiselen', *The Journal of Geography* 82/6 (1983), pp. 265–70, p. 267.
142 'Geography in the summer schools', p. 370.
143 Blouet, Brian W. and Stitcher, Teresa L., 'Survey of early geography teaching in state universities and land grant institutions', in B. W. Blouet (ed.), *The Origins of Academic Geography in the United States* (Hamden, CT: Archon Books, 1981), pp. 327–42; and Winsted, Huldah L., 'Geography in American universities', *The Journal of Geography* 10/10 (1912), pp. 309–16.
144 UCoB. *University of Colorado Bulletin*, 1915.
145 UCoB. *University of Colorado Bulletin*, 1915.
146 'Geography in the summer schools', p. 370.
147 Dodge, Stanley D., 'Ralph Hall Brown, 1898–1948', *Annals of the Association of American Geographers* 38/4 (1948), pp. 305–09; and Karan, Pradyumna P. and Mather, Cotton, *Leaders in American Geography*, vol. 2 (Mesilla: The New Mexico Geographical Society, 2000).
148 Barrows, Harlan H., 'Geography as human ecology', *Annals of the Association of American Geographers* 13/1 (1923), pp. 1–14; and McManis, Douglas R., 'Leading ladies at the AGS', *Geographical Review* 86/2 (1996), pp. 270–77.
149 Karan and Mather: *Leaders in American Geography*, p. 49.
150 AGSL. Fred E. Lukermann to John K. Wright, 19 May 1961.
151 WKU. *Normal Heights*, April 1917, p. 19.
152 'Meeting in Kentucky', *The Journal of Geography* 15/10 (1917), p. 336.
153 'Meeting in Kentucky', p. 336.
154 WKU. *College Heights Herald*, June 1929, p. 1.
155 WKU. *College Heights Herald*, June 1929, p. 1.

156 Harrison, Lowell H., *Western Kentucky University* (Lexington: University Press of Kentucky, 1987).

157 WKU. 'The Ellen Churchill Semple Geographical Society', 1931.

158 WKU. *The Talisman*, 1931, p. 132.

159 WKU. *The Talisman*, 1931, p. 132.

160 Jeffries, Ella, 'The dependence of the social sciences upon geographical principles', *The Journal of Geography* 26/6 (1926), pp. 228–36, p. 229.

161 Jeffries: 'Dependence of the social sciences', p. 229.

162 Keighren, Innes M., 'Giving voice to geography: popular lectures and the diffusion of knowledge', *Scottish Geographical Journal* 124/2 & 3 (2008), pp. 198–203.

163 Ríos-Font, Wadda C., *The Canon and the Archive: Configuring Literature in Modern Spain* (Lewisburg, PA: Bucknell University Press, 2004), p. 128.

CHAPTER 5

1 RGS, Correspondence Block 1881–1910. Semple to John S. Keltie, 6 March 1910.

2 Pattison, William D., 'Goode's proposal of 1902: an interpretation', *The Professional Geographer* 30/1 (1978), pp. 3–8.

3 Goode, J. Paul, 'The human response to the physical environment', *The Journal of Geography* 3/7 (1904), pp. 333–43, p. 342.

4 Pattison, William D., 'Rollin Salisbury and the establishment of geography at the University of Chicago', in B. W. Blouet (ed.), *The Origins of Academic Geography in the United States* (Hamden, CT: Archon Books, 1981), pp. 151–63, p. 157.

5 AGSL. Charles C. Colby to John K. Wright, 23 May 1961.

6 UCB, BANC MSS 77/170 c, Carton 4, Folder 3. Notebook of class notes, Chicago, 1909.

7 Whitbeck, Ray H., 'Thirty years of geography in the United States', *The Journal of Geography* 20/4 (1921), 121–28, p. 122.

8 Barrows, Harlan H., *Lectures on the Historical Geography of the United States as Given in 1933*, W. A. Koelsch (ed.) (Chicago: Department of Geography, University of Chcago, 1962), p. v; and Chappell, John C. Jr, 'Harlan Barrows and environmentalism', *Annals of the Association of American Geographers* 61/1 (1971), pp. 198–201.

9 University of Chicago, Special Collections Research Center (hereafter UC). *Circular of Information* 6/2 (1906), unpaginated.

10 Harris, Chauncy D., 'Geography at Chicago in the 1930s and 1940s', *Annals of the Association of American Geographers* 69/1 (1979), pp. 21–32, p. 23.

11 Colby, Charles C. and White, Gilbert F., 'Harlan H. Barrows, 1877–1960', *Annals of the Association of American Geographers* 51/4 (1961), pp. 395–400.

12 Bushong, Allen D., 'Geographers and their mentors: a genealogical view of American academic geography', in B. W. Blouet (ed.), *The Origins of Academic Geography in the United States* (Hamden, CT: Archon Books, 1981), pp. 193–219.

13 Rose, John K., 'Stephen Sargent Visher, 1887–1967', *Annals of the Association of American Geographers* 61/2 (1971), pp. 394–406, p. 394.

14 Rose: 'Stephen Sargent Visher', p. 395.

15 Colby, Charles C., 'Changing currents of geographic thought in America', *Annals of the Association of American Geographers* 26/1 (1936), pp. 1–37.

16 Raup, Hugh M., 'Trends in the development of geographic botany', *Annals of the Association of American Geographers* 32/4 (1942), pp. 319–54, p. 331.

17 Rose: 'Stephen Sargent Visher', p. 395.

18 University of South Dakota, Archives and Special Collections (hereafter USD). *Thirtieth Annual Catalogue of the University of South Dakota 1911–1912* (1912).

19 AGSL. Stephen S. Visher to John K. Wright, 25 March 1961.

20 Visher, Stephen S., 'Climatic influences', in G. Taylor (ed.), *Geography in the Twentieth Century* (London: Methuen, 1957), pp. 196–220.

21 Martin, Geoffrey J., 'Geography, geographers and Yale University, c. 1770–1970', in J. E. Harmon and T. J. Richards (eds), *Geography in New England* (New Britain, CT: New England/St Lawrence Valley Geographical Society, 1988), pp. 2–9.

22 Martin, Geoffrey J., *Ellsworth Huntington: His Life and Thought* (Hamden, CT: Archon Books, 1973).

23 Keltie, John S., 'Thirty years' progress in geographical education', *The Geographical Teacher* 7/38 (1914), pp. 215–27, p. 224.

24 PU, C0100, Box 115, Folder 12. Edward N. Bristol to Semple, 20 February 1912.

25 Yale University, Manuscripts and Archives (hereafter YU), MS 496D, Box 10, Folder 156. Semple to Millicent Todd, 19 December 1915.

26 UK, 46M139, Box 10. Ellsworth Huntington to Semple, 15 December 1931.

27 Roorbach, George B., 'The trend of modern geography', *Bulletin of the American Geographical Society* 46/11 (1914), pp. 801–16.

28 Roorbach: 'Trend of modern geography', p. 803.

29 Brigham, Albert P., 'Problems of geographic influence', *Annals of the Association of American Geographers* 5 (1915), pp. 3–25, p. 3; and Abrahams, Paul P., 'Academic geography in America: an overview', *Reviews in American History* 3/1 (1975), pp. 46–52.

30 Roorbach: 'Trend of modern geography', pp. 803–04.

31 Roorbach: 'Trend of modern geography', p. 803.

32 AGSL. George B. Cressey to John K. Wright, 6 April 1961; James, Preston E., 'George Babcock Cressey 1896–1963', *Geographical Review* 54/2 (1964), pp. 254–57; and James, Preston E. and Perejda, Andrew D., 'George Babcock Cressey 1896–1963', in T. W. Freeman (ed.), *Geographers: Biobibliographical Studies*, vol. 5 (London: Mansell, 1981), pp. 21–25.

33 Foscue, Edwin J., 'The life and works of Doctor Frank Carney', *Southwestern Social Science Quarterly* 16/3 (1935), pp. 51–59; Hubbard, George D., 'Sketch of the life and work of Dr Frank Carney', *The Ohio Journal of Science* 35/4 (1935), pp. 273–74; and Mahard, Richard H., 'A history of the department of geology and geography, Denison University, Granville, Ohio', *The Ohio Journal of Science* 79/1 (1979), pp. 18–21.

34 Brigham, Albert P., 'The School of Geography in the summer session of Cornell University', *Science* 18/455 (1903), pp. 380–81; and Foscue: 'Life and works of Doctor Frank Carney'.

35 Brigham, Albert P., 'Memoir of Ralph Stockman Tarr', *Annals of the Association of American Geographers* 3 (1913), pp. 93–98; and Engeln, Oscar D. von, 'Ralph Stockman Tarr', *Bulletin of the American Geographical Society* 44/4 (1912), pp. 283–85.

36 'Cornell summer school of geology and geography', *Bulletin of the American Geographical Society* 35/2 (1903), pp. 177–79; and Brigham: 'Summer session of Cornell University'.

37 See, for example, Carney, Frank, 'Geographic conditions in the early history of the Ohio country', *Bulletin of the Scientific Laboratories of Denison University* 16 (1911), pp. 403–23; Carney, Frank, 'Geographic influences in the development of Ohio', *Popular Science Monthly* 75 (1909), pp. 479–89; and Carney,

Frank, 'Springs as a geographic influence in humid climates', *Popular Science Monthly* 72 (1908), pp. 503–11.

38 Denison University, Archives and Special Collections (hereafter DU). *Denison University Annual Catalogue 1915–1916* (1916).

39 Bork, Kennard B., 'Kirtley Fletcher Mather's life in science and society', *The Ohio Journal of Science* 82/3 (1982), pp. 74–95, p. 77.

40 AGSL. George B. Cressey to John K. Wright, 6 April 1961.

41 Bork: 'Kirtley Fletcher Mather's life', p. 77.

42 Bork: 'Kirtley Fletcher Mather's life', p. 77.

43 Karan, Pradyumna P. and Mather, Cotton, *Leaders in American Geography*, vol. 2 (Mesilla: The New Mexico Geographical Society, 2000), p. 74.

44 Cressey, George B., 'Current trends in geography', *Science* 84/2179 (1936), pp. 311–14, p. 313; and Herman, Theodore, 'George Babcock Cressey, 1896–1963', *Annals of the Association of American Geographers* 55/2 (1965), pp. 360–64.

45 AGSL. George B. Cressey to John K. Wright, 6 April 1961.

46 AGSL. George B. Cressey to John K. Wright, 6 April 1961.

47 AGSL. George B. Cressey to John K. Wright, 6 April 1961.

48 AGSL. George B. Cressey to John K. Wright, 6 April 1961.

49 AGSL. George B. Cressey to John K. Wright, 6 April 1961. Capitalization in original.

50 Hubbard: 'Life and work of Dr Frank Carney', p. 274.

51 Bork, Kennard B., *Cracking Rocks and Defending Democracy: Kirtley Fletcher Mather, Scientist, Teacher, Social Activist, 1888–1978* (San Francisco: American Association for the Advancement of Science, Pacific Division, 1994).

52 UC, Rollin D Salisbury Papers, Box 5, Folder 14. James H. Hance to Salisbury, 9 January 1920.

53 Numbers, Ronald L. and Stephens, Lester D., 'Darwinism in the American south', in R. L. Numbers and J. Stenhouse (eds), *Disseminating Darwinism: The Role of Place, Race, Religion, and Gender* (Cambridge: Cambridge University Press, 1999), pp. 123–43.

54 Bork: 'Kirtley Fletcher Mather's life', p. 82.

55 Mather, Kirtley F., 'Evolution and religion', *The Scientific Monthly* 21/3 (1925), pp. 322–28.

56 Barbour, George B., 'John Lyon Rich, 1884–1956', *Annals of the Association of American Geographers* 48/2 (1958), pp. 174–77.

57 Fellmann, Jerome D., 'Development of geography at the University of Illinois' (paper presented at the Annual Meeting of the Association of American Geographers, Chicago, Illinois, 1995).

58 Fellmann, Jerome D., *Geography at Illinois: The Discipline and the Department, 1867–1974* (Urbana: Department of Geography, University of Illinois at Urbana-Champaign, 1974), p. 38.

59 Fellmann: *Geography at Illinois*, p. 38.

60 Rich, John L., 'Cultural features and the physiographic cycle', *Geographical Review* 4/4 (1917), pp. 297–308, p. 297.

61 Rich: 'Cultural features and the physiographic cycle', p. 306, p. 308.

62 Fellmann, 'Development of geography at the University of Illinois', unpaginated.

63 Bushong, Allen D., 'Ellen Churchill Semple 1863–1932', in T. W. Freeman (ed.), *Geographers: Biobibliographical Studies*, vol. 8 (London: Mansell, 1984), pp. 87–94, p. 92.

64 UC, Robert S. Platt Papers, Box 3, Folder 6. Platt to Walter M. Kollmorgen, 12 May 1956.

65   Beck, Joanna E., 'Environmental determinism in twentieth century American geography: reflections in the professional journals' (Ph.D. diss., University of California, 1985), p. 17.

66   UCB, BANC MSS 77/170 c, Carton 4, Folder 3. Notebook of class notes, Chicago, 1909; Kenzer, Martin S., 'Milieu and the "intellectual landscape": Carl O. Sauer's undergraduate heritage', *Annals of the Association of American Geographers* 75/2 (1985), pp. 258–70; and Leighly, John, 'Carl Ortwin Sauer, 1889–1975', *Annals of the Association of American Geographers* 66/3 (1976), pp. 337–48.

67   Sauer, Carl O., 'The education of a geographer', *Annals of the Association of American Geographers* 46/3 (1956), pp. 287–99, p. 288; and Sauer, Carl O., 'The fourth dimension of geography', *Annals of the Association of American Geographers* 64/2 (1974), pp. 189–92.

68   Speth, William W., *How it Came to Be: Carl O. Sauer, Franz Boas and the Meanings of Anthropogeography* (Ellensburg: Ephemera Press, 1999), p. 182.

69   'Annual general meeting', *The Geographical Journal* 141/3 (1975), pp. 519–24, p. 521.

70   Sauer, Carl O., *Seventeenth Century North America* (Berkeley: Turtle Island Foundation, 1980), p. 9.

71   The results of Sauer's study were published as Sauer, Carl O., 'Geography of the Upper Illinois Valley and history of development', *Illinois State Geological Survey* Bulletin no. 27 (1916).

72   Sauer: *Seventeenth Century North America*, p. 9.

73   Sauer: *Seventeenth Century North America*, p. 9.

74   UCB, BANC MSS 77/170 c, Box 18. Sauer to William W. Speth, 3 March 1972.

75   UC, Robert S. Platt Papers, Box 3, Folder 6. Platt to Walter M. Kollmorgen, 12 May 1956.

76   UCB, BANC MSS 77/170 c, Box 18. Sauer to William W. Speth, 3 March 1972.

77   Sauer: *Seventeenth Century North America*, p. 9.

78   Sauer: *Seventeenth Century North America*, pp. 9–10. Exclamation in original.

79   Sauer: *Seventeenth Century North America*, p. 10.

80   UCB, BANC MSS 77/170 c, Carton 4, Folder 24. Field notes (book 1), Missouri, Ozarks, February, 1914.

81   Quoted in Leighly, John, 'Carl Ortwin Sauer 1889–1975', in T. W. Freeman and P. Pinchemel (eds), *Geographers: Biobibliographical Studies*, vol. 2 (London: Mansell, 1978), pp. 99–108, p. 100.

82   UC, Robert S. Platt Papers, Box 3, Folder 6. Platt to Walter M. Kollmorgen, 12 May 1956.

83   Colby, Charles C., 'Narrative of five decades', in *A Half Century of Geography – What Next?* (Chicago: Department of Geography, University of Chicago, 1955), pp. 8–24, p. 11.

84   Colby, Charles C., 'Wellington Downing Jones, 1886–1957', *Annals of the Association of American Geographers* 50/1 (1960), pp. 51–54, p. 51; and Willis, Bailey, *A Yanqui in Patagonia* (Palo Alto, CA: Stanford University Press, 1948).

85   Platt, Robert S., 'Wellington Downing Jones', *Geographical Review* 48/2 (1958), pp. 285–87, p. 286.

86   UCB, BANC MSS 77/170 c, Box 11. Sauer to Richard Hartshorne, 22 June 1946.

87   Jones, Wellington D. and Sauer, Carl O., 'Outline for field work in geography', *Bulletin of the American Geographical Society* 47/7 (1915), pp. 520–25, p. 521.

88   Jones and Sauer: 'Outline for field work in geography', p. 524.

89   UC, Robert S. Platt Papers, Box 3, Folder 6. Platt to Walter M. Kollmorgen, 12 May 1956.
90   Leighly, John, 'What has happened to physical geography?', *Annals of the Association of American Geographers* 45/4 (1955), pp. 309–18, p. 315.
91   Leighly: 'Carl Ortwin Sauer, 1889–1975', p. 338.
92   Whitaker, J. Russell, 'Almon Ernest Parkins', *Annals of the Association of American Geographers* 31/1 (1941), pp. 46–50.
93   'Summer session courses in geography', *Geographical Review* 3/6 (1917), pp. 491–96.
94   HUA, HUG 4877.417. Charles C. Colby to Whittlesey, 4 February 1954.
95   Thoman, Richard S., 'Robert Swanton Platt 1891–1964', in T. W. Freeman and P. Pinchemel (eds), *Geographers: Biobibliographical Studies*, vol. 3 (London: Mansell, 1979), pp. 107–16.
96   Hartshorne, Richard, 'Robert S. Platt, 1891–1964', *Annals of the Association of American Geographers* 54/5 (1964), pp. 630–637.
97   UC, Robert S. Platt Papers, Box 7, Folder 6. Platt, Robert S., 'Being a geographer', *St George's Alumni Bulletin* XXX/1 (1947), unpaginated.
98   UC, Robert S. Platt Papers, Box 3, Folder 6. Platt to Walter M. Kollmorgen, 12 May 1956.
99   UC, Robert S. Platt Papers, Box 3, Folder 6. Platt to Walter M. Kollmorgen, 12 May 1956.
100 UC, Robert S. Platt Papers, Box 1, Folder 18. Notes from Barrows' 'Influences of Geography on American History', undated.
101 UC, Robert S. Platt Papers, Box 1, Folder 17. Unofficial transcript.
102 UC, Robert S. Platt Papers, Box 10, Folder 16. 'Changes in geographic thought', 19 March 1950.
103 UC, Robert S. Platt Papers, Box 10, Folder 16. 'Changes in geographic thought', 19 March 1950.
104 UC, Robert S. Platt Papers, Box 10, Folder 14. 'Can we avoid determinism?'.
105 AGSL. Robert S. Platt to John K. Wright, 3 April 1961.
106 Thoman: 'Robert Swanton Platt', p. 109.
107 Thoman: 'Robert Swanton Platt', p. 110.
108 Koelsch, William A., 'Derwent Stainthorpe Whittlesey 1890–1956', in P. H. Armstrong and G. J. Martin (eds), *Geographers: Biobibliographical Studies*, vol. 25 (London: Continuum, 2006), pp. 138–58.
109 UK, 46M139, Box 10. Derwent S. Whittlesey to Semple, 9 December 1931.
110 HUA, HUG 4877.412. Whittlesey to Semple, 2 February 1931.
111 Smith, Neil, '"Academic war over the field of geography": the elimination of geography at Harvard, 1947–1951', *Annals of the Association of American Geographers* 77/2 (1987), pp. 155–72, p. 162; and Ackerman, Edward A., 'Derwent Stainthorpe Whittlesey', *Geographical Review* 47/3 (1957), pp. 443–45.
112 HUA, HUG 4877.417. Whittlesey to Charles C. Colby, 26 January 1954.
113 James, Preston E. and Martin, Geoffrey J., *All Possible Worlds: A History of Geographical Ideas*, 2nd ed. (New York: John Wiley & Sons, 1981), p. 327.
114 Semple, Ellen C., 'The barrier boundary of the Mediterranean basin and its northern breaches as factors in history', *Annals of the Association of American Geographers* 5 (1915), pp. 27–59; and Semple, Ellen C., 'Pirate coasts of the Mediterranean Sea', *Geographical Review* 2/2 (1916), pp. 134–51.
115 Smith, Neil, *American Empire: Roosevelt's Geographer and the Prelude to Globalization* (Berkeley: University of California Press, 2003).
116 Wright, John K., 'The American Geographical Society: 1852–1952', *The Scientific Monthly* 74/3 (1952), pp. 121–31.

117 'War services of members of the Association of American Geographers', *Annals of the Association of American Geographers* 9 (1919), pp. 53–70.

118 Martin, Geoffrey J., *The Life and Thought of Isaiah Bowman* (Hamden, CT: Archon Books, 1980).

119 Martin, Geoffrey J., 'The emergence and development of geographic thought in New England', *Economic Geography* 74 (1998), pp. 1–13.

120 Martin, Geoffrey J., 'On Whittlesey, Bowman and Harvard', *Annals of the Association of American Geographers* 78/1 (1988), pp. 152–58.

121 PU, C0100, Box 115, Folder 12. Semple to Joseph Vogelius, 26 May 1911.

122 RGS, Correspondence Block 1881–1910. Semple to John S. Keltie, 18 May 1918.

123 Bronson, Judith C., 'Ellen Semple: contributions to the history of American geography' (Ph.D. diss., Saint Louis University, 1973), p. 110.

124 Crampton, Jeremy W., 'The cartographic calculation of space: race mapping and the Balkans at the Paris Peace Conference of 1919', *Social & Cultural Geography* 7/5 (2006), pp. 731–52; and Gelfand, Lawrence E., *The Inquiry: American Preparations for Peace, 1917–1919* (New Haven, CT: Yale University Press, 1963).

125 UK, 46M139, Box 2. Semple to Mrs Francis McVey, 24 May 1931.

126 James and Martin: *All Possible Worlds*, p. 346.

127 Bushong: 'Ellen Churchill Semple', p. 89.

128 De Bres, Karen J., 'An early frost: geography in Teachers College, Columbia and Columbia University, 1896–1942', *The Geographical Journal* 155/3 (1989), pp. 392–402, p. 394.

129 APS, Mss.B.B61, Series II. Courses of instruction in anthropology, 1915.

130 Mikesell, Marvin W., 'Geographic perspectives in anthropology', *Annals of the Association of American Geographers* 57/3 (1967), pp. 617–34, p. 627.

131 Trindell, Roger T., 'Franz Boas and American geography', *The Professional Geographer* 21/5 (1969), pp. 328–32, p. 329.

132 *Influences of Geographic Environment*, Northwestern University, Main Library, call number 551.4 S47 copy 2.

133 Rowley, Virginia M., *J. Russell Smith: Geographer, Educator, and Conservationist* (Philadelphia: University of Pennsylvania Press, 1964).

134 Martin, Geoffrey J., 'J. Russell Smith 1874–1966', in P. H. Armstrong and G. J. Martin (eds), *Geographers: Biobibliographical Studies*, vol. 21 (London: Mansell, 2001), pp. 97–113.

135 Barnes, Trevor J., '"In the beginning was economic geography" – a science studies approach to disciplinary history', *Progress in Human Geography* 25/4 (2001), pp. 521–44, p. 537.

136 APS, B Sm59. 'Journal of European trips, 1901–1903', 27 November 1901.

137 Freeman, Thomas W., 'Review of *J. Russell Smith: Geographer, Educator, Conservationist*, by Virginia M. Rowley', *The Geographical Journal* 132/1 (1966), p. 144.

138 Rowley: *J. Russell Smith*, p. 79.

139 PU, C0100, Box 115, Folder 12. Semple to Joseph Vogelius, 26 May 1911.

140 De Bres: 'Geography in Teachers College', p. 397.

141 De Bres: 'Geography in Teachers College', p. 397.

142 APS, B Sm59, Columbia University: geography course outlines. 'Geography 2 – Economic Geography, Columbia College', 1922.

143 Smith, J. Russell, 'Economic geography and its relation to economic theory and higher education', *Bulletin of the American Geographical Society* 39/8 (1907), pp. 472–81, p. 473.

144 Smith: 'Economic geography', p. 475.

145 Koelsch, William A., 'Wallace Atwood's "Great Geographical Institute"', *Annals of the Association of American Geographers* 70/4 (1980), pp. 567–82, p. 573.

146 Atwood, Wallace W., *The Clark Graduate School of Geography Our First Twenty-Five Years 1921–1946.* (Worcester, MA: Clark University, 1946).

147 Koelsch, William A., 'Geography at Clark: the first fifty years, 1921–1971', in J. E. Harmon and T. J. Richards (eds), *Geography in New England* (New Britain, CT: New England/St. Lawrence Valley Geographical Society, 1988), p. 41.

148 Colby, Charles C., 'Ellen Churchill Semple', *Annals of the Association of American Geographers* 23/4 (1933), pp. 229–40, p. 236.

149 Bronson, Judith C., 'A further note on sex discrimination and geography: the case of Ellen Churchill Semple', *The Professional Geographer* 27/1 (1975), pp. 111–12, p. 111.

150 Berman, Mildred, 'Sex discrimination and geography: the case of Ellen Churchill Semple', *The Professional Geographer* 26/1 (1974), pp. 8–11.

151 CU, Vertical file: 'Semple, Ellen Churchill'. Codicil III of the will of Ellen C. Semple, 1932.

152 CU, Vertical file: 'Semple, Ellen Churchill'. Codicil III of the will of Ellen C. Semple, 1932.

153 CU, Vertical file: 'Semple, Ellen Churchill'. Codicil III of the will of Ellen C. Semple, 1932.

154 Monk, Janice J., 'The women were always welcome at Clark', *Economic Geography* 74 (1998), pp. 14–30, p. 19.

155 Berman, Mildred, 'The geographical influence of Ellen Churchill Semple at Clark University' (paper presented at the Geography and legacy of Ellen Churchill Semple, Vassar College, Poughkeepsie, New York, 1992).

156 Harris, Chauncy D., 'Geographers in the U.S. government in Washington, DC, during World War II', *The Professional Geographer* 49/2 (1997), pp. 245–56.

157 Baugh, Ruth E., 'Ellen Churchill Semple, teacher', *The Monadnock* 6/2 (1932), p. 3.

158 Semple, Ellen C., 'The influence of geographic conditions upon ancient Mediterranean stock-raising', *Annals of the Association of American Geographers* 13 (1922), pp. 3–38.

159 Rowley: *J. Russell Smith*, p. 30.

160 Rowley: *J. Russell Smith*, p. 35.

161 Barrows, Harlan H., 'Geography as human ecology', *Annals of the Association of American Geographers* 13/1 (1923), pp. 1–14, p. 3.

162 James and Martin: *All Possible Worlds*, p. 319.

163 Rowley: *J. Russell Smith*, p. 35.

164 Leighly: 'Carl Ortwin Sauer, 1889–1975', p. 338.

165 Leighly, John, 'Drifting into geography in the Twenties', *Annals of the Association of American Geographers* 69/1 (1979), pp. 4–9.

166 Sauer, Carl O., 'Status and change in the rural Midwest – a retrospect', in W. A. Bladen and P. P. Karan (eds), *The Evolution of Geographic Thought in America: A Kentucky Root* (Dubuque, IA: Kendall/Hunt, 1983), pp. 115–22, p. 116.

167 Kenzer: 'Carl O. Sauer's undergraduate heritage', p. 267.

168 Sauer, Carl O., 'The survey method in geography and its objectives', *Annals of the Association of American Geographers* 14/1 (1924), pp. 17–33, p. 18.

169 Sauer: 'Survey method in geography', p. 18.

170 Sauer: 'Survey method in geography', p. 18.

171 Solot, Michael, 'Carl Sauer and cultural evolution', *Annals of the Association of American Geographers* 76/4 (1986), pp. 508–20; and Steward, Julian H., Gibson,

Ann J., and Rowe, John H., 'Alfred Louis Kroeber, 1876-1960', *American Anthropologist* 63/5 (1961), pp. 1038–87.

172 UCB, BANC MSS 77/170 c, Box 18. Sauer to William W. Speth, 3 March 1972.

173 UCB, BANC MSS 77/170 c, Box 11. Sauer to Richard Hartshorne, 22 June 1946.

174 Sauer, Carl O., 'The morphology of landscape', *University of California Publications in Geography* 2/2 (1925), pp. 19–54.

175 HUA, HUG 4877.412. Carl O. Sauer to Whittlesey, 25 September 1929.

176 Beck: 'Environmental determinism', p. 125.

177 CU, B4-18-11. Carl O. Sauer to Semple, 20 June 1926.

178 UCB, BANC MSS 77/170 c, Box 18. Semple to Sauer, 16 July 1918; and CU, B4-18-11. Carl O. Sauer to Semple, 20 June 1926.

179 Sauer, Carl O., 'Memorial of Ruliff S. Holway', *Annals of the Association of American Geographers* 19/1 (1929), pp. 64–65.

180 Dunbar, Gary S., *Geography in the University of California (Berkeley and Los Angeles) 1868–1941* (Marina del Ray, CA: DeVorss, 1981); and Stadtman, Verne A. (ed.), *The Centennial Record of the University of California* (Berkeley, CA: University of California Printing Department, 1967).

181 Speth, William W., 'Berkeley geography, 1923–33', in B. W. Blouet (ed.), *The Origins of Academic Geography in the United States* (Hamden, CT: Archon Books, 1981), pp. 221–44, p. 224.

182 UCB. *Annual Announcement of Courses of Instruction for the Academic Year 1910–11*(1910), pp. 113–115

183 UCB. *Annual Announcement of Courses of Instruction for the Academic Year 1918–19 (1918)*, p. 114.

184 AGSL. Clarence J. Glacken to John K. Wright, 11 May 1961; and John Leighly to John K. Wright, 29 March 1961.

185 Walker, H. Jesse, 'Richard Joel Russell 1895–1971', in T. W. Freeman and P. Pinchemel (eds), *Geographers: Biobibliographical Studies*, vol. 4 (London: Mansell, 1980), pp. 127–38, p. 128; and AGSL. Richard J. Russell to John K. Wright, 5 April 1961.

186 Kniffen, Fred B., 'Richard Joel Russell, 1895–1971', *Annals of the Association of American Geographers* 63/2 (1973), pp. 241–49; and McIntire, William G., 'Richard Joel Russell (1895–1971)', *Geographical Review* 63/2 (1973), pp. 276–79.

187 Walker, H. Jesse and Richardson, Miles E., 'Fred Bowerman Kniffen, 1900–1993', *Annals of the Association of American Geographers* 84/4 (1994), pp. 732–43.

188 McKee, Jesse O., 'Interview with Fred B. Kniffen', *The Mississippi Geographer* 4/1 (1976), pp. 5–6.

189 Walker and Richardson: 'Fred Bowerman Kniffen', p. 734.

190 Entrikin, J. Nicholas, 'Carl O. Sauer, philosopher in spite of himself', *Geographical Review* 74/4 (1984), pp. 387–408.

191 Mathewson, Kent and Shoemaker, Vincent J., 'Louisiana State University geography at seventy-five: "Berkeley on the bayou" and beyond', in J. O. Wheeler and S. D. Brunn (eds), *The Role of the South in the Making of American Geography: Centennial of the AAG, 2004* (Columbia, MD: Bellweather, 2004), pp. 245–67.

192 Mathewson, Kent and Shoemaker, Vincent J., 'Louisiana State University's Department of Geography and Anthropology: a selective history', *Southwestern Geographer* 1 (1997), pp. 62–84.

193 McIntire: 'Richard Joel Russell', p. 277.

194 AGSL. Gary S. Dunbar to John K. Wright, 24 May 1961.

195 AGSL. Gary S. Dunbar to John K. Wright, 24 May 1961.

196 Dangberg, Grace, *A Guide to the Life and Works of Frederick J. Teggart* (Reno, NV: Grace Dangberg Foundation, 1983); and Macpherson, Anne, 'Clarence James Glacken 1909–1989', in G. J. Martin (ed.), *Geographers: Biobibliographical Studies*, vol. 14 (London: Mansell, 1993), pp. 27–42.

197 AGSL. Clarence J. Glacken to John K. Wright, 11 May 1961.

198 AGSL. Clarence J. Glacken to John K. Wright, 11 May 1961. Underlining in original.

199 AGSL. Clarence J. Glacken to John K. Wright, 11 May 1961.

200 AGSL. Clarence J. Glacken to John K. Wright, 11 May 1961.

201 AGSL. Clarence J. Glacken to John K. Wright, 11 May 1961.

202 AGSL. Clarence J. Glacken to John K. Wright, 11 May 1961.

203 Macpherson: 'Clarence James Glacken', p. 33.

204 Glacken, Clarence J., *Traces on the Rhodian Shore: Nature and Culture in Western Thought From Ancient Times to the End of the Eighteenth Century* (Berkeley and Los Angeles: University of California Press, 1967).

205 UCB, CU-468, Box 7, Folder 32. Twentieth century.

206 Bruman, Henry J., Fetty, Archine, and Spencer, Joseph E., 'Ruth Emily Baugh, Geography: Los Angeles', *In Memoriam* (1976), pp. 10–12.

207 CU, B4-18-11. Wallace W. Atwood to Semple, 8 June 1925.

208 CU, B4-18-11. Semple to Wallace W. Atwood, 1 July 1925.

209 Martin, Geoffrey J., 'Preston E. James, 1899–1986', *Annals of the Association of American Geographers* 78/1 (1988), pp. 164–75.

210 Martin, Geoffrey J., 'Preston Everett James 1899–1986', in T. W. Freeman (ed.), *Geographers: Biobibliographical Studies*, vol. 11 (London: Mansell, 1987), pp. 63–70.

211 Quoted in Martin: 'Preston E. James', p. 165.

212 AGSL. Preston E. James to John K. Wright, 11 April 1961.

213 AGSL. Preston E. James to John K. Wright, 11 April 1961.

214 Quoted in Martin: 'Preston E. James', p. 166.

215 James, Preston E., 'The Blackstone Valley: a study in chorography in Southern New England', *Annals of the Association of American Geographers* 19/2 (1929), pp. 67–109; and James, Preston E., 'A geographic reconnaissance of Trinidad', *Economic Geography* 3/1 (1927), pp. 87–109.

216 James: 'Blackstone Valley', p. 117.

217 James and Martin: *All Possible Worlds*, p. 327.

218 James and Martin: *All Possible Worlds*, p. 327.

219 AGSL. Preston E. James to John K. Wright, 11 April 1961.

220 AGSL. Preston E. James to John K. Wright, 11 April 1961.

221 AGSL. Philip W. Porter to John K. Wright, 30 March 1961.

222 AGSL. Ruth E. Baugh to John K. Wright, 24 May 1961.

223 AGSL. Ruth E. Baugh to John K. Wright, 24 May 1961.

224 Spencer, Joseph E., 'A geographer west of the Sierra Nevada', *Annals of the Association of American Geographers* 69/1 (1979), pp. 46–52, p. 46.

225 AGSL. Ruth E. Baugh to John K. Wright, 24 May 1961.

226 Nelson, Howard J., 'J. E. Spencer, 1907–1984', *Annals of the Association of American Geographers* 75/4 (1985), pp. 595–603.

227 Spencer: 'Geographer west of the Sierra Nevada', p. 47.

228 Spencer: 'Geographer west of the Sierra Nevada', p. 47.

229 Nelson: 'J. E. Spencer', p. 596.

230 Bruman, Fetty, and Spencer: 'Ruth Emily Baugh', p. 11.

231 'Dr Semple completes work on the Mediterranean region', *The Monadnock* 6/1 (1931), p. 10; and Bushong, Allen D., 'Ruth Emily Baugh (1889–1975): a cen-

tennial assessment' (paper presented at the Annual Meeting of the Association of American Geographers, Baltimore, Maryland, 1989).

232 Nash, Peter H., Guelke, Leonard, and Preston, Richard E., *Abstract Thoughts, Concrete Solutions: Essays in Honour of Peter Nash* (Waterloo, Ontario: Department of Geography, University of Waterloo, 1987); and Walker, H. Jesse, 'Evelyn Lord Pruitt, 1918–2000', *Annals of the Association of American Geographers* 96/2 (2006), pp. 432–39.

233 Nash, Guelke, and Preston: *Essays in Honour of Peter Nash*, p. 280.

234 AGSL. George Tatham to John K. Wright, undated.

235 Barnes, Carleton P., 'W. Elmer Ekblaw, 1882–1949', *Annals of the Association of American Geographers* 39/4 (1949), p. 295.

236 Ekblaw, W. Elmer, 'The material response of the polar Eskimo to their far Arctic environment', *Annals of the Association of American Geographers* 17/4 (1927), pp. 147–98; and Ekblaw, W. Elmer, 'The material response of the polar Eskimo to their far Arctic environment (continued)', *Annals of the Association of American Geographers* 18/1 (1928), pp. 1–24.

237 Beck: 'Environmental determinism', p. 82.

238 AGSL. George Tatham to John K. Wright, undated.

239 AGLS. Albert S. Carlson to John K. Wright, undated; Van H. English to John K. Wright, undated; Wilma B. Fairchild to John K. Wright, 26 March 1961; Stephen B. Jones to John K. Wright, 27 March 1961; and Walter W. Ristow to John K. Wright, 18 April 1961.

240 AGSL. George Tatham to John K. Wright, undated.

241 Ekblaw, W. Elmer, 'Review of *The Geography of the Mediterranean Region*, by Ellen C. Semple', *Economic Geography* 8/1 (1932), pp. 104–05, p. 104.

242 Prunty, Merle C., 'Clark in the early 1940s', *Annals of the Association of American Geographers* 69/1 (1979), pp. 42–45, p. 45.

243 Prunty: 'Clark in the early 1940s', p. 45.

244 Bushong: 'Geographers and their mentors'; and Koelsch, William A., 'East and Midwest in American academic geography: two prosopographic notes', *The Professional Geographer* 53/1 (2001), pp. 97–105.

245 AGSL. George Tatham to John K. Wright, undated.

246 AGSL. George Tatham to John K. Wright, undated.

247 AGSL. George Tatham to John K. Wright, undated.

248 AGSL. Albert S. Carlson to John K. Wright, undated.

249 AGSL. Albert S. Carlson to John K. Wright, undated.

250 Simpson, Robert E. and Huke, Robert E., 'One view of geography at Dartmouth College', in J. E. Harmon and T. J. Richards (eds), *Geography in New England* (New Britain, CT: New England/St. Lawrence Valley Geographical Society, 1988), pp. 10–24.

251 AGSL. Stephen B. Jones to John K. Wright, 27 March 1961.

252 AGSL. Stephen B. Jones to John K. Wright, 27 March 1961.

253 Harris, Chauncy D., 'Stephen Barr Jones, 1903–1984', *Annals of the Association of American Geographers* 75/2 (1985), pp. 271–76.

254 AGSL. Stephen B. Jones to John K. Wright, 27 March 1961.

255 AGSL. Stephen B. Jones to John K. Wright, 27 March 1961.

256 AGSL. Stephen B. Jones to John K. Wright, 27 March 1961.

257 Said, Edward W., *The World, the Text, and the Critic* (London: Vintage, 1991), p. 227.

258 AGLS. John Leighly to John K. Wright, 29 March 1961.

259 Wright, John K., 'Miss Semple's "Influences of geographic environment" notes towards a bibliobiography', *Geographical Review* 52/3 (1962), pp. 346–61, pp. 349–50.

260 Wright: 'Notes towards a bibliobiography', p. 349.

261 Bowen, Emrys G., 'Geography in the University of Wales, 1918–1948', in R. W. Steel (ed.), *British Geography 1918–1945* (Cambridge: Cambridge University Press, 1987).

262 Watson, J. Wreford, 'The sociological aspect of geography', in G. Taylor (ed.), *Geography in the Twentieth Century* (London: Methuen, 1957), pp. 463–99, p. 472.

263 Watson: 'Sociological aspect of geography', p. 471.

264 AGSL. Emyr E. Evans to John K. Wright, 14 April 1961.

265 Buchanan, Ronald H., 'Emyr Estyn Evans 1905–1989', in P. H. Armstrong and G. J. Martin (eds), *Geographers: Biobibliographical Studies*, vol. 25 (London: Continuum, 2006), pp. 13–33, p. 14.

266 AGSL. Emyr E. Evans to John K. Wright, 14 April 1961.

267 Carter, Harold, 'Emrys G. Bowen, 1900–1983', *Transactions of the Institute of British Geographers* 9/3 (1984), pp. 374–80; Fleure, Herbert J., 'Harold John Edward Peake, 1867–1946', *Man* 47 (1947), pp. 48–50; and Williams, Michael, 'The creation of humanised landscapes', in R. Johnston and M. Williams (eds), *A Century of British Geography* (Oxford: Oxford University Press for the British Academy, 2003), pp. 167–212, p. 172.

268 Freeman, Thomas W., *A Hundred Years of Geography* (London: Duckworth, 1961), p. 180.

269 Freeman: *Modern British Geography*, p. 110.

270 Buchanan, Ronald H., Jones, Emrys, and McCourt, Desmond, *Man and His Habitat: Essays Presented to Emyr Estyn Evans* (London: Routledge and Keegan Paul, 1971).

271 AGSL. Emyr E. Evans to John K. Wright, 14 April 1961.

272 Keltie, John S., *The Position of Geography in British Universities*, W. L. G. Joerg (ed.) (New York: Oxford University Press, 1921); and Scargill, David I., 'The RGS and the foundations of geography at Oxford', *The Geographical Journal* 142/3 (1976), pp. 438–61.

273 Keltie: *Position of Geography in British Universities*, pp. 9–10.

274 AGSL. John N. L. Baker to John K. Wright, 30 April 1961.

275 AGSL. John N. L. Baker to John K. Wright, 30 April 1961.

276 AGSL. John N. L. Baker to John K. Wright, 30 April 1961.

277 Baker, John N. L., 'The history of geography in Oxford', in J. N. L. Baker (ed.), *The History of Geography: Papers by J. N. L. Baker* (Oxford: Basil Blackwell, 1963), pp. 119–29.

278 Steel, Robert W., 'John Norman Leonard Baker 1893–1971', in G. J. Martin (ed.), *Geographers: Biobibliographical Studies*, vol. 16 (London: Mansell, 1995), p. 5.

279 AGSL. John N. L. Baker to John K. Wright, 30 April 1961.

280 AGSL. John N. L. Baker to John K. Wright, 30 April 1961.

281 AGSL. John N. L. Baker to John K. Wright, 30 April 1961.

282 *Influences of Geographic Environment*, University of Oxford, Geography and the Environment Library, call number M 59a, p. 245, p. 299.

283 *Influences of Geographic Environment*, University of Oxford, Geography and the Environment Library, call number M 59a, p. 413.

284 CU, B4-18-11. 'Miss Semple's list. European geographers', unpaginated.

285 Dritsas, Lawrence, 'Expeditionary science: knowledge between field and metropolis' (paper presented at the Geographies of Nineteenth-Century Science, University of Edinburgh, 2007); and Fleure, Herbert J., 'Alfred Cort Haddon 1855–1940', *Obituary Notices of Fellows of the Royal Society* 3/9 (1941), pp. 448–65.

286 Cambridge University Library (hereafter CUL), Department of Manuscripts and University Archives, CUR 28.18. *Cambridge University Register*, 14 June 1909, unpaginated; and Scargill, David I., 'The RGS and the foundations of geography at Oxford', *The Geographical Journal* 142/3 (1976), pp. 438–61.

287 CUL, Min.V.31., Minute Book, Board of Geographical Studies. 19 November 1915 and 23 May 1918; and Davis, William M., 'Geography at Cambridge University, England', *The Journal of Geography* 19/6 (1920), pp. 207–10.

288 Baker, Alan R. H., 'Henry Clifford Darby 1909–1992', *Transactions of the Institute of British Geographers* 17/4 (1992), pp. 495–501; and Darby, Henry C., 'Academic geography in Britain: 1918–1946', *Transactions of the Institute of British Geographers* 8/1 (1983), pp. 14–26.

289 Darby: 'Academic geography in Britain', p. 16.

290 AGSL. Henry C. Darby to John K. Wright, 17 April 1961.

291 Meller, Helen, *Patrick Geddes: Social Evolutionist and City Planner* (London: Routledge, 1993), p. 37.

292 Tatham, George, 'Environmentalism and possibilism', in G. Taylor (ed.), *Geography in the Twentieth Century* (London: Methuen, 1957), pp. 128–62, p. 151.

293 Aspects of possibilism are described in Livingstone, David N., *The Geographical Tradition: Episodes in the History of a Contested Enterprise* (Oxford: Blackwell, 1992), pp. 68–69; and Tatham, 'Environmentalism and possibilism'.

294 AGSL. Henry C. Darby to John K. Wright, 17 April 1961; and Dickinson, Robert E., *The Makers of Modern Geography* (New York: Frederick A. Praeger, 1969), p. 193.

295 AGSL. Henry C. Darby to John K. Wright, 17 April 1961.

296 AGSL. Henry C. Darby to John K. Wright, 17 April 1961.

297 Darby, Henry C., *The Theory and Practice of Geography: An Inaugural Lecture Delivered at Liverpool on 7 February 1946* (Liverpool: University Press of Liverpool, 1946); and Freeman, Thomas W., 'Percy Maude Roxby 1880–1947', in T. W. Freeman (ed.), *Geographers: Biobibliographical Studies*, vol. 5 (London: Mansell, 1981), pp. 109–16.

298 AGSL. George Tatham to John K. Wright, undated.

299 AGSL. George Tatham to John K. Wright, undated.

300 Crone, Gerald R., 'British geography in the twentieth century', *The Geographical Journal* 130/2 (1964), pp. 197–220, p. 202; and Clout, Hugh, 'Place description, regional geography and area studies', in R. J. Johnston and M. Williams (eds), *A Century of British Geography* (Oxford: Oxford University Press, 2003), pp. 247–74.

301 Freeman: 'Percy Maude Roxby', p. 113, p. 110.

302 Freeman: 'Percy Maude Roxby', p. 114.

303 AGSL. George Tatham to John K. Wright, undated.

304 Warkentin, John, 'George Tatham 1907–1987', *The Canadian Geographer* 31/4 (1987), p. 381.

305 Dosse, François, *New History in France: The Triumph of the Annales*, tr. P. V. Conroy (Urbana: University of Illinois Press, 1994).

306 Ward, R. Gerard, 'Oskar Hermann Khristian Spate (1911–2000)', *Australian Geographical Studies* 39/2 (2001), pp. 253–55.

307 AGSL. Oskar H. K. Spate to John K. Wright, 4 April 1961.

308 AGSL. Oskar H. K. Spate to John K. Wright, 4 April 1961.

309 AGSL. Oskar H. K. Spate to John K. Wright, 4 April 1961.

310 AGSL. Oskar H. K. Spate to John K. Wright, 4 April 1961; for a description of probabilism, see Martin, A. F., 'The necessity for determinism: a metaphysical problem confronting geographers', *Transactions and Papers (Institute of British Geographers)* 17 (1951), pp. 1–11.

311 AGSL. Oskar H. K. Spate to John K. Wright, 4 April 1961.

312 AGSL. Oskar H. K. Spate to John K. Wright, 4 April 1961.

313 Withers, Charles W. J., 'Working with old maps: tracing the reception and legacy of Blaeu's 1654 *Atlas Novus*', *Scottish Geographical Journal* 121/3 (2005), pp. 297–310.

314 *Influences of Geographic Environment*, University of Oxford, Geography and the Environment Library, call number M 59a, p. 413.

315 *Influences of Geographic Environment*, University of Birmingham, Main Library, shelfmark GB 95/S copy 1, p. 620.

316 Wise, Michael J., 'Becoming a geographer around the second world war', *Progress in Human Geography* 25/1 (2001), pp. 112–21, p. 113.

317 'Robert Henry Kinvig', *Transactions of the Institute of British Geographers* 49 (1970), pp. 519–24.

318 Wise: 'Becoming a geographer', p. 113.

319 Wise: 'Becoming a geographer', p. 113.

320 *Influences of Geographic Environment*, University of Birmingham, Main Library, shelfmark GB 95/S copy 2, p. 19.

321 *Influences of Geographic Environment*, University of California at Berkeley, Doe Memorial Library, Gardner Main Stacks, call number GF31.S5 copy 3, p. 186.

322 *Influences of Geographic Environment*, University of Chicago, Joseph Regenstein Library, call number GF31.S4 copy 4, p. ix; and *Influences of Geographic Environment*, Queen's University Belfast, Main Library, shelfmark GF51/SEMP copy 1, p. 33.

323 *Influences of Geographic Environment*, University of Sheffield, Main Library, shelfmark B 910.01 (S), p. 189.

324 *Influences of Geographic Environment*, University of Oxford, Geography and the Environment Library, shelfmark M 59a, p. 474.

325 Barnes, Harry E., *History and Social Intelligence* (New York: Alfred A. Knopf, Inc., 1926), p. 64.

326 RGS, Charles F. A. Close Papers, LBR.MSS.2d (Vol. 3). Unstead, John F., 'Regional geography and its future development', unpaginated.

327 Edinburgh University Library, Special Collections (hereafter EUL). *The Edinburgh University Calendar 1925–1926*, unpaginated; and AGSL. J. Wreford Watson to John K. Wright, 2 June 1961.

328 RGS, Hugh Robert Mill Collection, Folder 7. Semple to Mill, 18 January 1931.

329 AGSL. J. Wreford Watson to John K. Wright, 2 June 1961.

330 Semple, Ellen C., 'Domestic and municipal waterworks in ancient Mediterranean lands', *Geographical Review* 21/3 (1931), pp. 466–74.

331 Bladen and Karan: *Evolution of Geographic Thought*, p. 46.

332 UK, 46M139, Box 10. J. Paul Goode to Semple, 5 February 1932.

333 UK, 46M139, Box 10. Fay-Cooper Cole to Semple, 24 March 1932.

334 UK, 46M139, Box 2. Semple to Frank L. McVey, 28 March 1932.

335 Koelsch, William A., 'Academic geography, American style: an institutional perspective', in G. S. Dunbar (ed.), *Geography: Discipline, Profession and Subject Since 1870* (Dordrecht: Kluwer Academic, 2001), pp. 245–79.

336 AGSL. George Tatham to John K. Wright, undated.

337 AGSL. Albert S. Carlson to John K. Wright, undated.

338 Livingstone, David N., 'Mobilising science: writers, readers and the geographies of meaning' (paper presented at the Geographies of Nineteenth-Century Science, University of Edinburgh, 2007), unpaginated.

339 Graff, Gerald, 'Is there a text in this class?', in H. A. Vesser (ed.), *The Stanley Fish Reader* (Malden, MA and Oxford: Blackwell, 1999), pp. 38–54, p. 38.

CHAPTER 6

1   Gelfand, Lawrence E., 'Ellen Churchill Semple: her geographical approach to American history', *The Journal of Geography* 53/1 (1954), pp. 30–37, p. 39.
2   These quotations are taken, in reverse order, from: Taylor, Eva G. R., 'Review of *The Geographic Background of Greek and Roman History*, by Max Cary', *The Geographical Journal* 114/1–3 (1949), pp. 83–84, p. 83; Crone, Gerald R., 'British geography in the twentieth century', *The Geographical Journal* 130/2 (1964), pp. 197–220, p. 202; and AGSL. John F. Hart to John K. Wright, 11 May 1961.
3   Leighly, John, 'Review of *The Geographic Factor: Its Rôle in Life and Civilization*, by Ray H. Whitbeck and Olive J. Thomas', *American Anthropologist* 35/4 (1933), pp. 766–67, p. 766.
4   McEwan, Cheryl, 'Cutting power lines within the palace? Countering paternity and eurocentrism in the "geographical tradition"', *Transactions of the Institute of British Geographers* 23/3 (1998), pp. 371–84, p. 11.
5   Domosh, Mona, 'Toward a feminist historiography of geography', *Transactions of the Institute of British Geographers* 16/1 (1991), pp. 95–104; Maddrell, Avril M. C., *Complex Locations: Women's Geographical Work in the UK 1850–1970* (Chichester: Wiley-Blackwell, 2009); Maddrell, Avril M. C., 'Scientific discourse and the geographical work of Marion Newbigin', *Scottish Geographical Magazine* 113/1 (1997), pp. 33–41; Monk, Janice J., 'The women were always welcome at Clark', *Economic Geography* 74 (1998), pp. 14–30; Monk, Janice J., 'Women, gender, and the histories of American geography', *Annals of the Association of American Geographers* 94/1 (2004), pp. 1–22; Monk, Janice J., 'Women's worlds at the American Geographical Society', *Geographical Review* 93/2 (2004), pp. 237–57; Monk, Janice J., Fortuijn, Joos D., and Raleigh, Clionadh, 'The representation of women in academic geography: contexts, climate and curricula', *Journal of Geography in Higher Education* 28/1 (2004), pp. 83–90; Rubin, Barbara, '"Women in geography" revisited: present status, new options', *The Professional Geographer* 31/2 (1979), pp. 125–34; Zelinsky, Wilbur, 'The strange case of the missing female geographer', *The Professional Geographer* 25/2 (1973), pp. 101–05; and Zelinsky, Wilbur, 'Women in geography: a brief factual account', *The Professional Geographer* 25/2 (1973), pp. 151–65.
6   For a discussion of geography's historiographical practices see Livingstone, David N., *The Geographical Tradition: Episodes in the History of a Contested Enterprise* (Oxford: Blackwell, 1992).
7   See, for example, Murphy, David T., '"A sum of the most wonderful things": *Raum*, geopolitics and the German tradition of environmental determinism, 1900–1933', *History of European Ideas* 25/3 (1999), pp. 121–33.
8   Radcliffe, Sarah A., Watson, Elizabeth E., Simmons, Ian, Fernández-Armesto, Felipe, and Sluyter, Andrew, 'Forum: Environmentalist thinking and/in geography', *Progress in Human Geography* 34/1 (2009), pp. 98–116; and Sluyter, Andrew, 'Neo-environmental determinism, intellectual damage control, and nature/society science', *Antipode* 35/4 (2003), pp. 813–17.
9   Colby, Charles C., 'Ellen Churchill Semple', *Annals of the Association of American Geographers* 23/4 (1933), pp. 229–40, p. 232.
10 Roorbach, George B., 'Review of *Influences of Geographic Environment*, by Ellen C. Semple', *Annals of the American Academy of Political and Social Science* 41 (1912), pp. 350–51, p. 350.

11 Livingstone, David N., 'Text, talk and testimony: geographical reflections on scientific habits. An afterword', *British Journal for the History of Science* 38/1 (2005), pp. 93–100, p. 99.

12 Nersessian, Nancy J., 'Opening the black box: cognitive science and history of science', *Osiris* 10 (1995), pp. 194–211, p. 208.

13 Price, Leah, 'Review of *In Another Country: Colonialism, Culture, and the English Novel in India*, by Priya Joshi', *Victorian Studies* 45/2 (2003), pp. 333–34, p. 334.

14 Barnes, Trevor J., 'Performing economic geography: two men, two books, and a cast of thousands', *Environment and Planning A* 34/3 (2002), pp. 487–512; and Mayhew, Robert J., 'The character of English geography *c.* 1660–1800: a textual approach', *Journal of Historical Geography* 24/4 (1998), pp. 385–412.

15 Secord, James A., 'Knowledge in transit', *Isis* 95/4 (2004), pp. 654–72, p. 664.

Bibliography

PRIMARY SOURCES

American Geographical Society Archives

Isaiah Bowman Collection
Folder 'Semple, Ellen Churchill 1920–32'. Semple to John K. Wright, 1 February [1922].

American Geographical Society Library

Replies to a questionnaire relating to Ellen Churchill Semple's Influences of Geographic Environment, *1961*
John N. L. Baker to John K. Wright, 30 April 1961.
Ruth E. Baugh to John K. Wright, 24 May 1961.
Albert S. Carlson to John K. Wright, undated.
Charles C. Colby to John K. Wright, 23 May 1961.
George B. Cressey to John K. Wright, 6 April 1961.
Henry C. Darby to John K. Wright, 17 April 1961.
Gary S. Dunbar to John K. Wright, 24 May 1961.
Van H. English to John K. Wright, undated.
Emyr E. Evans to John K. Wright, 14 April 1961.
Wilma B. Fairchild to John K. Wright, 26 March 1961.
Herbert J. Fleure to John K. Wright, 31 May 1961.
Herbert J. Fleure to John K. Wright, 3 June 1961.
Clarence J. Glacken to John K. Wright, 11 May 1961.
John F. Hart to John K. Wright, 11 May 1961.
Preston E. James to John K. Wright, 11 April 1961.
Stephen B. Jones to John K. Wright, 27 March 1961.
John Leighly to John K. Wright, 29 March 1961.
Fred E. Luckermann to John K. Wright, 19 May 1961.
Robert S. Platt to John K. Wright, 3 April 1961.
Philip W. Porter to John K. Wright, 30 March 1961.
Walter W. Ristow to John K. Wright, 18 April 1961.

Richard J. Russell to John K. Wright, 5 April 1961.
Oskar H. K. Spate to John K. Wright, 4 April 1961.
George Tatham to John K. Wright, undated.
Stephen S. Visher to John K. Wright, 25 March 1961.
J. Wreford Watson to John K. Wright, 2 June 1961.

American Philosophical Society Library

Franz Boas Collections, Mss.B.B61
Series II. Courses of instruction in anthropology, 1915.
Series IV. 'Franz Boas: his work as described by some of his contemporaries', undated typescript.

Joseph Russell Smith Papers, B Sm59
Columbia University: geography course outlines. 'Geography 2 – Economic Geography, Columbia College' typescript, 1922.
'Journal of European Trips, 1901–1903'.

Cambridge University Library
(Department of Manuscripts and University Archives)

CUR 28.18. *Cambridge University Register*, 14 June 1909, unpaginated.
Min.V.31., Minute Book, Board of Geographical Studies. 19 November 1915 and 23 May 1918.

Clark University (Archives and Special Collections)

Vertical file: 'Semple, Ellen Churchill'. Codicil III of the will of Ellen C. Semple, 1932.

Wallace W. Atwood Papers
B4-18-11. Semple to Wallace W. Atwood, 29 August 1922.
B4-18-11. 'Miss Semple's list. European geographers', undated [1923].
B4-18-11. Wallace W. Atwood to Semple, 8 June 1925.
B4-18-11. Semple to Wallace W. Atwood, 1 July 1925.
B4-18-11. Carl O. Sauer to Semple, 20 June 1926.
B4-18-11. Duren J. H. Ward to Atwood, 2 July 1932.

Columbia University (Rare Book and Manuscript Library)

Edward V. D. Robinson Papers, Ms Coll Robinson, E. V.
Box 3, Binder 5, Section 10. Davis R. Dewey to Robinson 15 December 1911.

Denison University (Archives and Special Collections)

Denison University Annual Catalogue 1915–1916 (1916).

Edinburgh University Library (Special Collections)

The Edinburgh University Calendar 1925–1926.

Harvard University Archives

Derwent Stainthorpe Whittlesey Papers
HUG 4877.410. Charles C. Colby to Whittlesey, 18 June 1932.
HUG 4877.412. Semple to Whittlesey, 4 April 1929.
HUG 4877.412. Carl O. Sauer to Whittlesey, 25 September 1929.
HUG 4877.412. Whittlesey to Semple, 2 February 1931.
HUG 4877.417. Whittlesey to Charles C. Colby, 26 January 1954.
HUG 4877.417. Charles C. Colby to Whittlesey, 4 February 1954.

National Library of Wales

Herbert John Fleure, A1982/3 1977152
Item 7. Reminiscences, undated.

The Newberry Library

Floyd Dell Papers, 1908–1969, Midwest MS Dell
Box 11, Folder 426. Semple to Dell, [1911].
Box 11, Folder 426. Semple to Dell, 18 November 1911.

Oxford University Archives (Bodleian Library)

School of Geography
GE 4/1. Scrapbook of biennial vacation courses 1908–1922.
GE 5. Account book of biennial vacation courses 1912–1935.

Princeton University (Department of Rare Books and Special Collections)

Archives of Henry Holt, C0100
Box 115, Folder 12. Semple to Edward N. Bristol, 13 March 1911.
Box 115, Folder 12. Semple to Joseph Vogelius, 26 May 1911.
Box 115, Folder 12. Semple to Edward N. Bristol, 4 January 1912.
Box 115, Folder 12. Semple to Edward N. Bristol, 12 February 1912.

Box 115, Folder 12. Edward N. Bristol to Semple, 20 February 1912.
Box 115, Folder 12. Edward N. Bristol to Semple, 11 March 1912.
Box 115, Folder 12. Semple to Treasurer, 8 November 1912.
Box 115, Folder 12. Treasurer to Semple, 16 November 1912.
Box 115, Folder 12. Treasurer to United States Trust Company, 10 April 1933.

Royal Geographical Society

Additional Papers, Box 93/2. Cutting from *The Guardian*, 10 July 1982.
Additional Papers, Box 93/2. Freshfield, Douglas W., 'Memorandum on the conduct of affairs by the Council of the Royal Geographical Society', 1894.
Correspondence Block 1881–1910. Semple to John S. Keltie, 12 August 1905.
Correspondence Block 1881–1910. Semple to John S. Keltie, 2 September 1905.
Correspondence Block 1881–1910. Semple to John S. Keltie, 21 April 1907.
Correspondence Block 1881–1910. Semple to John S, Keltie, 8 March 1908.
Correspondence Block 1881–1910. Semple to John S. Keltie, 6 March 1910.
Correspondence Block 1881–1910. John S. Keltie to Semple, 16 March 1910.
Correspondence Block 1881–1910. Semple to John S. Keltie, 18 May 1918.
Correspondence Block 1881–1911. Clements Markham to John S. Keltie, 10 September 1911.
Correspondence Block 1881–1911. George G. Chisholm to John S. Keltie, 18 September 1911.
Correspondence Block 1881–1911. Hugh Robert Mill to John S. Keltie, 23 September 1911.
Correspondence Block 1911–1920. Semple to John S. Keltie, 2 April 1911.
Correspondence Block 1911–1920. John S. Keltie to Semple, 22 April 1911.
Correspondence Block 1911–1920. Semple to John S. Keltie, 19 August 1911.
Correspondence Block 1911–1920. John S. Keltie to Semple, 6 September 1911.
Correspondence Block 1911–1920. Semple to John S. Keltie, 11 November 1911.
Correspondence Block 1911–1920. John S. Keltie to Semple, 11 December 1911.
Correspondence Block 1911–1920. Semple to John S. Keltie, 23 January 1912.
Correspondence Block 1911–1920. Semple to John S. Keltie, 5 July 1912.
Correspondence Block 1911–1920. Semple to John S. Keltie, 30 October 1912.
Correspondence Block 1921–1930. Semple to Hugh R. Mill, 19 December 1924.
Correspondence Block 1921–1930. Annual Awards 1922, Paper proposing Ellen Churchill Semple, A.M., LL.D.

Charles F. A. Close Papers
LBR.MSS.2d (vol. 3) Unstead, John F., 'Regional geography and its future development', unpaginated.

Hugh Robert Mill Collection
Folder 7. Semple to Mill, 18 January 1931.

University of California at Berkeley (The Bancroft Library)

Annual Announcement of Courses of Instruction for the Academic Year 1910–11 (1910).
Annual Announcement of Courses of Instruction for the Academic Year 1918–19 (1918).

Clarence J. Glacken Papers, CU-468
Box 1, Folder 2. Glacken to Thomas R. Smith, 19 April 1963.
Box 7, Folder 32. Twentieth century.
Box 7, Folder 34. Mss. on ideas, 1974.

Carl O. Sauer Papers, BANC MSS 77/170 c
Box 11. Sauer to Richard Hartshorne, 22 June 1946.
Box 12. Preston E. James to Sauer, 25 January 1972.
Box 18. Sauer to William W. Speth, 3 March 1972.
Box 18. Semple to Sauer, 16 July 1918.
Carton 4, Folder 3. Notebook of class notes, Chicago, 1909.
Carton 4, Folder 24. Field notes (book 1), Missouri, Ozarks, February, 1914.

University of Chicago (Special Collections Research Center)

Circular of Information 6/2 (1906).

Robert S. Platt Papers
Box 1, Folder 17. Unofficial transcript, undated.
Box 1, Folder 18. Notes from Barrows' 'Influences of Geography on American History', undated.
Box 3, Folder 6. Platt to Walter M. Kollmorgen, 12 May 1956.
Box 7, Folder 6. Platt, Robert S., 'Being a geographer', *St George's Alumni Bulletin* XXX/1 (1947), unpaginated.
Box 10, Folder 14. 'Can we avoid determinism?', undated.
Box 10, Folder 16. 'Changes in geographic thought', 19 March 1950.

Rollin D Salisbury Papers
Box 5, Folder 14. James H. Hance to Salisbury, 9 January 1920.

University of Colorado at Boulder (Archives)

University of Colorado Bulletin 12/2 (1915).

University of Kentucky (Special Collections and Digital Programs)

Ellen Churchill Semple Papers, 46M139
Box 2. Baugh, Ruth E, 'Ellen Churchill Semple, the great lady of American geography', 30 August 1961.
Box 2. Semple to Mrs Francis McVey, 24 May 1931.
Box 2. Semple to Frank L. McVey, 28 March 1932.
Box 10. Scrapbook, 1895–1932.
Box 10. Scrapbook, 1895–1932. Geographical Circle programme, undated.
Box 10. Derwent S. Whittlesey to Semple, 9 December 1931.
Box 10. Ellsworth Huntington to Semple, 15 December 1931.
Box 10. J. Paul Goode to Semple, 5 February 1932.
Box 10. Fay-Cooper Cole to Semple, 24 March 1932.

University of South Dakota (Archives and Special Collections)

Thirtieth Annual Catalogue of the University of South Dakota 1911–1912 (1912).
Thirty-first Annual Catalogue of the University of South Dakota 1912–1913 (1913).

Wellesley College Archives

Wellesley College Course Catalog, 1914–1915.
Wellesley College News, 3 December 1914.
1DB/1899–1966. Report of the Department of Geology and Geography, 1914–1915.

Western Kentucky University (Museum and Library)

College Heights Herald, June 1929.
'The Ellen Churchill Semple Geographical Society', 1931.
Normal Heights, April 1917.
The Talisman, 1931.

Yale University (Manuscripts and Archives)

Millicent Todd Bingham Papers, MS 496D
Box 10, Folder 156. Semple to Millicent Todd Bingham, 19 December 1915.

Newspapers and periodicals

The Aberdeen Free Press
American Library Association Booklist (Chicago)
The Bookseller (London)
Boston Evening Transcript (Massachusetts)

Boston Herald (Massachusetts)
Chicago Evening Post (Illinois)
The Continent
The Courier-Journal (Louisville, Kentucky)
Daily Express (London)
Daily-Picayune (New Orleans, Louisiana)
Dundee Advertiser
The Evening Post (New York City, New York)
The Guardian (London)
The Hartford Courant (Connecticut)
Irish Times (Dublin)
Louisville Herald (Kentucky)
The Louisville Times (Kentucky)
The Manchester Guardian
The Morning Post (London)
The Nation (New York City, New York)
Newark News (New Jersey)
The Outlook (New York City, New York)
The Oxford Times
The Post-Standard (Syracuse, New York)
Providence Daily Journal (Rhode Island)
The Publishers Weekly (New York City, New York)
Review of Reviews (New York City, New York)
The Saturday Review (London)
The Scotsman (Edinburgh)
Springfield Daily Republican (Massachusetts)
The Sun (New York City, New York)
The Times (London)
The Washington Post (District of Columbia)

Library copies of *Influences*

American Geographical Society Library, call number GF31 .S5.
Northwestern University, Main Library, call number 551.4 S47 copy 2.
Queen's University Belfast, Main Library, shelfmark GF51/SEMP copy 1.
University of Birmingham, Main Library, shelfmark GB 95/S copy 1.
University of Birmingham, Main Library, shelfmark GB 95/S copy 2.
University of California at Berkeley, Doe Memorial Library, Gardner Main Stacks, call number GF31.S5 copy 3.
University of Chicago, Joseph Regenstein Library, call number GF31.S4 copy 4.
University of Sheffield, Main Library, shelfmark B 910.01 (S).
University of Oxford, Geography and the Environment Library, call number M 59a.

SECONDARY SOURCES

Abrahams, Paul P., 'Academic geography in America: an overview', *Reviews in American History* 3/1 (1975), pp. 46–52.

Ackerman, Edward A., 'Derwent Stainthorpe Whittlesey', *Geographical Review* 47/3 (1957), pp. 443–45.

Almagià, Roberto, 'Review of *Influences of Geographic Environment*, by Ellen C. Semple', *Bollettino della Società Geographica Italiana* 5/1 (1912), pp. 550–54.

'An alpine section in a woman's club', *Bulletin of the American Geographical Society* 40/9 (1908), p. 553.

Anderson, Monica, *Women and the Politics of Travel, 1870–1914* (Cranbury, NJ: Fairleigh Dickinson University Press, 2006).

'Annual general meeting', *The Geographical Journal* 141/3 (1975), pp. 519–24.

Atwood, Wallace W., 'An appreciation of Ellen Churchill Semple, 1863–1932', *The Journal of Geography* 31/6 (1932), p. 267.

——, *The Clark Graduate School of Geography Our First Twenty-Five Years 1921–1946.* (Worcester, MA: Clark University, 1946).

Baigent, Elizabeth, 'Herbertson, Andrew John (1865–1915)', in H. C. G. Matthew and B. Harrison (eds), *Oxford Dictionary of National Biography* (Oxford: Oxford University Press, 2004).

Baker, Alan R. H., 'Henry Clifford Darby 1909–1992', *Transactions of the Institute of British Geographers* 17/4 (1992), pp. 495–501.

Baker, John N. L., 'The history of geography in Oxford', in J. N. L. Baker (ed.), *The History of Geography: Papers by J. N. L. Baker* (Oxford: Basil Blackwell, 1963), pp. 119–29.

Bannister, Robert C., *Sociology and Scientism: The American Quest for Objectivity, 1880–1940* (Chapel Hill: University of North Carolina Press, 1987).

Barbour, George B., 'John Lyon Rich, 1884–1956', *Annals of the Association of American Geographers* 48/2 (1958), pp. 174–77.

Barnes, Carleton P., 'W. Elmer Ekblaw, 1882–1949', *Annals of the Association of American Geographers* 39/4 (1949), p. 295.

Barnes, Harry E., *History and Social Intelligence* (New York: Alfred A. Knopf, Inc., 1926).

——, 'Some contributions of sociology to modern political theory', *The American Political Science Review* 15/4 (1921), pp. 487–533.

Barnes, Trevor J., '"In the beginning was economic geography" – a science studies approach to disciplinary history', *Progress in Human Geography* 25/4 (2001), pp. 521–44.

——, 'Performing economic geography: two men, two books, and a cast of thousands', *Environment and Planning A* 34/3 (2002), pp. 487–512.

Barnett, Clive, 'Awakening the dead: who needs the history of geography?', *Transactions of the Institute of British Geographers* 20/3 (1995), pp. 417–19.

Barrows, Harlan H., 'The department of geography: thirty years ago and now', *The University Record* 19/3 (1933), pp. 197–99.

——, 'Geography as human ecology', *Annals of the Association of American Geographers* 13/1 (1923), pp. 1–14.

——, *Lectures on the Historical Geography of the United States as Given in 1933*, W. A. Koelsch (ed.) (Chicago: Department of Geography, University of Chicago, 1962).

Barton, Thomas F. and Karan, Pradyumna P., *Leaders in American Geography*, vol. 1 (Mesilla: The New Mexico Geographical Society, 1992).

Bassin, Mark, 'Friedrich Ratzel 1844–1904', in T. W. Freeman (ed.), *Geographers: Biobibliographical Studies*, vol. 11 (London: Mansell, 1987), pp. 123–32.

——, 'History and philosophy of geography', *Progress in Human Geography* 21/4 (1997), pp. 563–72.

——, 'Reply: reductionism or redux? or the convolutions of contextualism', *Annals of the Association of American Geographers* 83/1 (1993), pp. 163–66.

——, 'Studying ourselves: history and philosophy of geography', *Progress in Human Geography* 24/3 (2000), pp. 475–87.

Baugh, Ruth E., 'Ellen Churchill Semple, teacher', *The Monadnock* 6/2 (1932), p. 3.

Beck, Hanno, 'Moritz Wagner als geograph', *Erdkunde* 7 (1953), pp. 125–28.

Beck, Joanna E., 'Environmental determinism in twentieth century American geography: reflections in the professional journals' (Ph.D. diss., University of California, 1985).

Beckinsale, Robert P., 'W. M. Davis and American geography: 1880–1934', in B. W. Blouet (ed.), *The Origins of Academic Geography in the United States* (Hamden, CT: Archon Books, 1981), pp. 107–22.

Beckit, Henry O., 'The United States National Parks', *The Geographical Journal* 42/4 (1913), pp. 333–42.

Bell, Morag, 'Reshaping boundaries: international ethics and environmental consciousness in the early twentieth century', *Transactions of the Institute of British Geographers* 23/2 (1998), pp. 151–75.

Bell, Morag and McEwan, Cheryl, 'The admission of women fellows to the Royal Geographical Society, 1892–1914; the controversy and the outcome', *The Geographical Journal* 162/3 (1996), pp. 295–312.

Ben-Itto, Hadassa, *The Lie That Wouldn't Die:* The Protocols of the Elders of Zion (London: Valentine Mitchell, 2005).

Benson, Lee, 'Achille Loria's influence on American economic thought: including his contributions to the frontier hypothesis', *Agricultural History* 24/4 (1950), pp. 182–99.

——, *Turner and Beard: American Historical Writing Reconsidered* (Glencoe, IL: The Free Press, 1960).

Berman, Jessica S., *Modernist Fiction, Cosmopolitanism, and the Politics of Community* (Cambridge: Cambridge University Press, 2001).

Berman, Mildred, 'The geographical influence of Ellen Churchill Semple at Clark University' (paper presented at the Geography and legacy of Ellen Churchill Semple, Vassar College, Poughkeepsie, New York, 1992).

——, 'Sex discrimination and geography: the case of Ellen Churchill Semple', *The Professional Geographer* 26/1 (1974), pp. 8–11.

Bingham, Millicent T., 'Ellen Churchill Semple, geographer', *Vassar Quarterly*, July (1932).

Bird, Isabella L., *Unbeaten Tracks in Japan: An Account of Travels in the Interior, Including Visits to the Aborigines of Yezo and the Shrines of Nikkô and Isé* (London: John Murray, 1880).

Birkett, Dea, *Spinsters Abroad: Victorian Lady Explorers* (Oxford: Basil Blackwell, 1989).

Bladen, Wilford A., 'Nathaniel Southgate Shaler and early American geography', in W. A. Bladen and P. P. Karan (eds), *The Evolution of Geographic Thought in America: A Kentucky Root* (Dubuque, IA: Kendall/Hunt, 1983), pp. 13–27.

Bladen, Wilford A. and Karan, Pradyumna P. (eds), *The Evolution of Geographic Thought in America: A Kentucky Root* (Dubuque, IA: Kendall/Hunt, 1983).

Blair, Ann, 'An early modernist's perspective', *Isis* 95/3 (2004), pp. 420–30.

——, 'Reading strategies for coping with information overload ca. 1550–1700', *Journal of the History of Ideas* 64/1 (2003), pp. 11–28

Block, Robert H., 'Frederick Jackson Turner and American geography', *Annals of the Association of American Geographers* 70/1 (1980), pp. 31–42.

Blouet, Brian W. and Stitcher, Teresa L., 'Survey of early geography teaching in state universities and land grant institutions', in B. W. Blouet (ed.), *The Origins of Academic Geography in the United States* (Hamden, CT: Archon Books, 1981), pp. 327–42.

Boas, Franz, 'The study of geography', *Science* 9/210 (1887), pp. 137–41.

Boatright, Mody C., 'The myth of frontier individualism', *Southwestern Social Science Quarterly* 22 (1941), pp. 14–32.

Bogue, Allan G., '"Not by bread alone": the emergence of the Wisconsin idea and the departure of Frederick Jackson Turner', *Wisconsin Magazine of History* 2002, pp. 10–23.

Bork, Kennard B., *Cracking Rocks and Defending Democracy: Kirtley Fletcher Mather, Scientist, Teacher, Social Activist, 1888–1978* (San Francisco: American Association for the Advancement of Science, Pacific Division, 1994).

——, 'Kirtley Fletcher Mather's life in science and society', *The Ohio Journal of Science* 82/3 (1982), pp. 74–95.

Bowen, Emrys G., 'Geography in the University of Wales, 1918–1948', in R. W. Steel (ed.), *British Geography 1918–1945* (Cambridge: Cambridge University Press, 1987).

Bowen, Emrys G., Carter, Harold, and Taylor, James A. (eds), *Geography at Aberystwyth: Essays Written on the Occasion of the Departmental Jubilee 1917–18 – 1967–68* (Cardiff: University of Wales Press, 1968).

Brandeis, Adele, 'Ellen Semple as I remember her', *Courier Journal*, 8 January 1963.

Brewer, William M., 'The historiography of Frederick Jackson Turner', *The Journal of Negro History* 44/3 (1959), pp. 240–59.

Brigham, Albert P., 'Memoir of Ralph Stockman Tarr', *Annals of the Association of American Geographers* 3 (1913), pp. 93–98.

——, 'Notes on the Transcontinental Excursion of the American Geographical Society', *The Journal of Geography* 6/4 (1913), pp. 155–58.

——, 'Problems of geographic influence', *Annals of the Association of American Geographers* 5 (1915), pp. 3–25.

——, 'The School of Geography in the summer session of Cornell University', *Science* 18/455 (1903), pp. 380–81.

Broek, Jan O. M., *Compass of Geography* (Columbus, OH: Charles E. Merril Books, 1966).

——, *Geography: Its Scope and Spirit* (Columbus, OH: Charles E. Merrill Books, 1965).

Bronson, Judith C., 'Ellen Semple: contributions to the history of American geography' (Ph.D. diss., Saint Louis University, 1973).

——, 'A further note on sex discrimination and geography: the case of Ellen Churchill Semple', *The Professional Geographer* 27/1 (1975), pp. 111–12.

Bruman, Henry J., Fetty, Archine, and Spencer, Joseph E., 'Ruth Emily Baugh, Geography: Los Angeles', *In Memoriam* (1976), pp. 10–12.

Brush, Stephen G., 'The reception of Mendeleev's periodic law in America and Britain', *Isis* 87/4 (1996), pp. 595–628.

Buchanan, Ronald H., 'Emyr Estyn Evans 1905–1989', in P. H. Armstrong and G. J. Martin (eds), *Geographers: Biobibliographical Studies*, vol. 25 (London: Continuum, 2006), pp. 13–33.

Buchanan, Ronald H., Jones, Emrys, and McCourt, Desmond, *Man and His Habitat: Essays Presented to Emyr Estyn Evans* (London: Routledge and Keegan Paul, 1971).

Buettner-Janusch, John, 'Boas and Mason: particularism versus generalization', *American Anthropologist* 59/2 (1957), pp. 318–24.

Bulletin of Vassar College: Alumnae Biographical Register Issue (Poughkeepsie, NY: Vassar College, 1939).

Burton, David H., 'Theodore Roosevelt's Social Darwinism and views on imperialism', *Journal of the History of Ideas* 26/1 (1965), pp. 103–18.

Bushong, Allen D., 'Ellen Churchill Semple 1863–1932', in T. W. Freeman (ed.), *Geographers: Biobibliographical Studies*, vol. 8 (London: Mansell, 1984), pp. 87–94.

——, 'Geographers and their mentors: a genealogical view of American academic geography', in B. W. Blouet (ed.), *The Origins of Academic Geography in the United States* (Hamden, CT: Archon Books, 1981), pp. 193–219.

——, 'Ruth Emily Baugh (1889–1975): a centennial assessment' (paper presented at the Annual Meeting of the Association of American Geographers, Baltimore, Maryland, 1989).

Butcher, Philip (ed.), *The William Stanley Braithwaite Reader* (Ann Arbor: University of Michigan Press, 1972).

Buttimer, Anne, *Society and Milieu in the French Geographic Tradition* (Chicago: Rand McNally, 1971).

Buttmann, Günther, *Friedrich Ratzel: Leben und Werk eines deutschen Geographen 1844–1904* (Stuttgart: Wissenschaftliche Verlagsgesellschaft, 1977).

Cahnman, Werner J., 'Methods of geopolitics', *Social Forces* 21/2 (1942), pp. 147–54.

Campbell, John A. and Livingstone, David N., 'Neo-Lamarckism and the development of geography in the United States and Great Britain', *Transactions of the Institute of British Geographers* 8/3 (1983), pp. 267–94.

Carney, Frank, 'Geographic conditions in the early history of the Ohio country', *Bulletin of the Scientific Laboratories of Denison University* 16 (1911), pp. 403–23.

——, 'Geographic influences in the development of Ohio', *The Journal of Geography* 9/7 (1911), pp. 169–74.

——, 'Geographic influences in the development of Ohio', *Popular Science Monthly* 75 (1909), pp. 479–89.

——, 'Springs as a geographic influence in humid climates', *Popular Science Monthly* 72 (1908), pp. 503–11.

Carter, C. C. and McGregor, C., 'Long vacation course at the School of Geography, Oxford: the work of the course', *The Geographical Teacher* 1 (1901–1902), pp. 172–74.

Carter, Harold, 'Emrys G. Bowen, 1900–1983', *Transactions of the Institute of British Geographers* 9/3 (1984), pp. 374–80.

Chamberlain, Houston S., *Die Grundlagen des neunzehnten Jahrhunderts*, 2 vols. (Munich: F. Bruckmann, 1899).

——, *The Foundations of the Nineteenth Century*, tr. J. Lees, 2 vols. (New York: John Lane, 1911).

Chamberlain, James F., 'Geographic influences in the development of California', *The Journal of Geography* 9/10 (1911), pp. 253–61.

Chamberlin, Joseph E., The Boston Transcript: *A History of its First Hundred Years* (Boston: Houghton Mifflin, 1930).

Chappell, John C. Jr, 'Harlan Barrows and environmentalism', *Annals of the Association of American Geographers* 61/1 (1971), pp. 198–201.

Chisholm, George G., 'The meaning and scope of geography', *Scottish Geographical Magazine* 24 (1908), pp. 561–75.

——, 'Miss Semple on the influences of geographical environment. Review of *Influences of Geographic Environment* by Ellen C. Semple', *The Geographical Journal* 39/1 (1912), pp. 31–37.

——, 'Some recent contributions to geography', *Scottish Geographical Magazine* 27 (1911), pp. 561–73.

Chorley, Richard J., Dunn, Antony J., and Beckinsale, Robert P., *The History of the Study of Landforms or the Development of Geomorphology, Volume 2: The Life and Work of William Morris Davis* (London: Methuen, 1973).

Circular of Information for the Year 1906–1907. Chicago: University of Chicago Press, 1906.

Clark, Brett and Foster, John B., 'George Perkins Marsh and the transformation of the earth: an introduction to Marsh's *Man and Nature*', *Organization & Environment* 15/2 (2002), pp. 164–69.

Close, Charles F., 'The position of geography', *The Geographical Journal* 38/4 (1911), pp. 404–13.

Clout, Hugh, 'Place description, regional geography and area studies', in R. J. Johnston and M. Williams (eds), *A Century of British Geography* (Oxford: Oxford University Press, 2003), pp. 247–74.

Colby, Charles C., 'Changing currents of geographic thought in America', *Annals of the Association of American Geographers* 26/1 (1936), pp. 1–37.

——, 'Ellen Churchill Semple', *Annals of the Association of American Geographers* 23/4 (1933), pp. 229–40.

——, 'Narrative of five decades', in *A Half Century of Geography – What Next?* (Chicago: Department of Geography, University of Chicago, 1955), pp. 8–24.

——, 'Wellington Downing Jones, 1886–1957', *Annals of the Association of American Geographers* 50/1 (1960), pp. 51–54.

Colby, Charles C. and White, Gilbert F., 'Harlan H. Barrows, 1877–1960', *Annals of the Association of American Geographers* 51/4 (1961), pp. 395–400.

Coleman, William, 'Science and symbol in the Turner frontier hypothesis', *The American Historical Review* 72/1 (1966), pp. 22–49.

Conlin, Michael F., 'The popular and scientific reception of the Foucault Pendulum in the United States', *Isis* 90/2 (1999), pp. 181–204.

'Cornell summer school of geology and geography', *Bulletin of the American Geographical Society* 35/2 (1903), pp. 177–79.

Crampton, Jeremy W., 'The cartographic calculation of space: race mapping and the Balkans at the Paris Peace Conference of 1919', *Social & Cultural Geography* 7/5 (2006), pp. 731–52.

Cressey, George B., 'Current trends in geography', *Science* 84/2179 (1936), pp. 311–14.

Croly, Jane C., *The History of the Woman's Club Movement in America* (New York: Henry G. Allen & Co., 1898).

Crone, Gerald R., 'British geography in the twentieth century', *The Geographical Journal* 130/2 (1964), pp. 197–220.

——, *Modern Geographers: An Outline of Progress in Geography Since 1800 A.D.* (London: Royal Geographical Society, 1951).

'The Cullum Geographical Medal presented to Miss Ellen Churchill Semple', *Bulletin of the American Geographical Society* 46/5 (1914), p. 364.

Curzon, George N., Gowland, William, Lyde, Lionel W., and Semple, Ellen C., 'Influences of geographic conditions upon Japanese agriculture – discussion', *The Geographical Journal* 40/6 (1912), pp. 603–07.

Dabney, Thomas E., *One Hundred Great Years: The Story of the* Times-Picayune *From its Founding to 1940* (Baton Rouge: Louisiana State University Press, 1944).

Dagg, Anne I., *The Feminine Gaze: A Canadian Compendium of Non-Fiction Women Authors and Their Books, 1836–1945* (Waterloo, ON: Wilfrid Laurier University Press, 2001).

Daingerfield, Henderson, 'Social settlement and educational work in the Kentucky mountains', *Journal of Social Science* 39 (1901), pp. 176–89.

Dangberg, Grace, *A Guide to the Life and Works of Frederick J. Teggart* (Reno, NV: Grace Dangberg Foundation, 1983).

Darby, Henry C., 'Academic geography in Britain: 1918–1946', *Transactions of the Institute of British Geographers* 8/1 (1983), pp. 14–26.

——, *The Theory and Practice of Geography: An Inaugural Lecture Delivered at Liverpool on 7 February 1946* (Liverpool: University Press of Liverpool, 1946).

Darwin, Leonard and Lyons, Henry G., 'Notes on Moroccan geography – discussion', *The Geographical Journal* 41/3 (1913), pp. 237–39.

Daston, Lorraine, 'Taking note(s)', *Isis* 95/3 (2004), pp. 443–48.

Davis, William M., 'Current notes of physiography', *Science* 14/351 (1901), pp. 457–59.

——, 'Geography at Cambridge University, England', *The Journal of Geography* 19/6 (1920), pp. 207–10.

Davis, William M. and Dodge, Richard E., 'The *Journal of School Geography*', *Science* 5/118 (1897), pp. 551–53.

De Bres, Karen J., 'An early frost: geography in Teachers College, Columbia and Columbia University, 1896–1942', *The Geographical Journal* 155/3 (1989), pp. 392–402.

de Souza, Anthony R., 'Talks with teachers: Elizabeth Eiselen', *The Journal of Geography* 82/6 (1983), pp. 265–70.

Dickinson, Robert E., *The Makers of Modern Geography* (New York: Frederick A. Praeger, 1969).

——, *Regional Concept: The Anglo-American Leaders* (London: Routledge and Kegan Paul, 1976).

'Dissertations in geography accepted by universities in the United States for the degree of Ph.D. as of May, 1935', *Annals of the Association of American Geographers* 25/4 (1935), pp. 211–37.

Dodge, Richard E., 'Edward Van Dyke Robinson', *Annals of the Association of American Geographers* 6 (1916), p. 120.

——, 'Man and his geographic environment', *The Journal of Geography* 8/8 (1910), pp. 179–87.

——, 'The social function of geography', *Journal of School Geography* 2/9 (1898), pp. 328–36.

Dodge, Stanley D., 'Ralph Hall Brown, 1898–1948', *Annals of the Association of American Geographers* 38/4 (1948), pp. 305–09.

Domosh, Mona, 'Beyond the frontiers of geographical knowledge', *Transactions of the Institute of British Geographers* 16/4 (1991), pp. 488–89.

——, 'Toward a feminist historiography of geography', *Transactions of the Institute of British Geographers* 16/1 (1991), pp. 95–104.

Dosse, François, *New History in France: The Triumph of the Annales*, tr. P. V. Conroy (Urbana: University of Illinois Press, 1994).

'Dr. Semple completes work on the Mediterranean region', *The Monadnock* 6/1 (1931), p. 10.

Dritsas, Lawrence, 'Expeditionary science: knowledge between field and metropolis' (paper presented at the Geographies of Nineteenth-Century Science, University of Edinburgh, 2007).

Dryer, Charles R., 'A century of geographic education in the United States', *Annals of the Association of American Geographers* 14/3 (1924), pp. 117–49.

——, 'Geographic influences in the development of India', *The Journal of Geography* 9/1 (1910), pp. 17–22.

Dumenil, Lynn, 'Love, literature, and politics in the machine age. Review of *Floyd Dell: The Life and Times of an American Rebel*, by Douglas Clayton', *Reviews in American History* 23/4 (1995), pp. 699–703.

Dunbar, Gary S., 'Credentialism and careerism in American geography, 1890–1915', in B. W. Blouet (ed.), *The Origins of Academic Geography in the United States* (Hamden, CT: Archon Books, 1981), pp. 71–88.

——, *Geography in the University of California (Berkeley and Los Angeles) 1868–1941* (Marina del Ray, CA: DeVorss, 1981).

Eisenstein, Elizabeth L., *The Printing Press as an Agent of Change: Communications and Cultural Transformations in Early-Modern Europe* (Cambridge: Cambridge University Press, 1979).

Ekblaw, W. Elmer, 'The material response of the polar Eskimo to their far Arctic environment', *Annals of the Association of American Geographers* 17/4 (1927), pp. 147–98.

——, 'The material response of the polar Eskimo to their far Arctic environment (continued)', *Annals of the Association of American Geographers* 18/1 (1928), pp. 1–24.

——, 'Review of *The Geography of the Mediterranean Region*, by Ellen C. Semple', *Economic Geography* 8/1 (1932), pp. 104–05.

Elder, Eleanor S., 'Women in early geology', *Journal of Geological Education* 30/5 (1982), pp. 287–93.

Elder, John, *Pilgrimage to Vallombrosa: From Vermont to Italy in the Footsteps of George Perkins Marsh* (Charlottesville: University of Virginia Press, 2006).

Engeln, Oscar D. von, 'Ralph Stockman Tarr', *Bulletin of the American Geographical Society* 44/4 (1912), pp. 283–85.

Entrikin, J. Nicholas., 'Carl O. Sauer, philosopher in spite of himself', *Geographical Review* 74/4 (1984), pp. 387–408.

Fawcett, Charles B., 'Recent losses to British geography', *Geographical Review* 37/3 (1947), pp. 505–07.

Febvre, Lucien and Martin, Henri-Jean, *The Coming of the Book: The Impact of Printing 1450–1800*, tr. D. Gerard (London: NLB, 1976).

Fellmann, Jerome D., 'Development of geography at the University of Illinois' (paper presented at the Annual Meeting of the Association of American Geographers, Chicago, Illinois, 1995).

——, *Geography at Illinois: The Discipline and the Department, 1867–1974* (Urbana: Department of Geography, University of Illinois at Urbana-Champaign, 1974).

Fellmann, Jerome D., 'Myth and reality in the origin of American economic geography', *Annals of the Association of American Geographers* 76/3 (1986), pp. 313–30.

Finnegan, Diarmid A., '"A free and liberal tone": scientific speech and intellectual culture in mid-Victorian Edinburgh' (paper presented at the Geographies of Nineteenth-Century Science, University of Edinburgh, 2007).

——, 'Natural history societies in late Victorian Scotland and the pursuit of local civic science', *British Journal for the History of Science* 38/1 (2005), pp. 53–72.

Fish, Stanley, *Is There a Text in This Class? The Authority of Interpretive Communities* (Cambridge, MA: Harvard University Press, 1980).

Fitzpatrick, Ellen F., *History's Memory: Writing America's Past, 1880–1980* (Cambridge, MA: Harvard University Press, 2004).

Fleure, Herbert J., 'Alfred Cort Haddon 1855–1940', *Obituary Notices of Fellows of the Royal Society* 3/9 (1941), pp. 448–65.

——, 'Harold John Edward Peake, 1867–1946', *Man* 47 (1947), pp. 48–50.

——, 'The later developments in Herbertson's thought: a study in the application of Darwin's ideas', *Geography* 37 (1952), pp. 97–103.

——, 'Recollections of A. J. Herbertson', *Geography* 50 (1965), pp. 348–49.

——, 'Review of *Influences of Geographic Environment*, by Ellen C. Semple', *The Geographical Teacher* 7/35 (1913), pp. 65–68.

Fleure, Herbert J. and James, T. C., 'Geographical distribution of anthropological types in Wales', *The Journal of the Royal Anthropological Institute of Great Britain and Ireland* 46 (1916), pp. 35–153.

Foscue, Edwin J., 'The life and works of Doctor Frank Carney', *Southwestern Social Science Quarterly* 16/3 (1935), pp. 51–59.

Foster, Alice, 'The new department in its setting', in *A Half Century of Geography – What Next?* (Chicago: Department of Geography, University of Chicago, 1955), pp. 1–7.

Frasca-Spada, Marina and Jardine, Nick (eds), *Books and the Sciences in History* (Cambridge: Cambridge University Press, 2000).

Freeman, Thomas W., 'Herbert John Fleure 1877–1969', in T. W. Freeman (ed.), *Geographers: Biobibliographical Studies*, vol. 11 (London: Mansell, 1987), pp. 35–51.

———, *A History of Modern British Geography* (London: Longman, 1980).

———, *A Hundred Years of Geography* (London: Duckworth, 1961).

———, 'The Manchester and Royal Scottish Geographical Societies', *The Geographical Journal* 150/1 (1984), pp. 55–62.

———, 'Percy Maude Roxby 1880–1947', in T. W. Freeman (ed.), *Geographers: Biobibliographical Studies*, vol. 5 (London: Mansell, 1981), pp. 109–16.

———, 'Review of *J. Russell Smith: Geographer, Educator, Conservationist*, by Virginia M. Rowley', *The Geographical Journal* 132/1 (1966), p. 144.

Frenkel, Stephen, 'Geography, empire, and environmental determinism', *Geographical Review* 82/2 (1992), pp. 143–53.

Freund, Rudolf, 'Turner's theory of social evolution', *Agricultural History* 19/2 (1945), 78–87.

Fyfe, Aileen, 'The reception of William Paley's *Natural Theology* in the University of Cambridge', *British Journal for the History of Science* 30/3 (1997), pp. 321–35.

———, *Science and Salvation: Evangelical Popular Science Publishing in Victorian Britain* (Chicago and London: University of Chicago Press, 2004).

Gage, William L., *The Life of Carl Ritter: Late Professor of Geography in the University of Berlin* (New York: Charles Scribner and Co, 1867).

Garnett, Alice, 'Herbert John Fleure: 1877–1969', *Biographical Memoirs of Fellows of the Royal Society* 16 (1970), pp. 253–78.

Geddes, Patrick, 'The influence of geographical conditions on social development', *The Geographical Journal* 12/6 (1898), pp. 580–86.

Gelfand, Lawrence E., 'Ellen Churchill Semple: her geographical approach to American history', *The Journal of Geography* 53/1 (1954), pp. 30–37.

———, *The Inquiry: American Preparations for Peace, 1917–1919* (New Haven, CT: Yale University Press, 1963).

Genthe, Martha K., 'Comment on Colonel Close's address on the purpose and position of geography', *Bulletin of the American Geographical Society* 44/1 (1912), pp. 27–38.

Genthe, Martha K., and Semple, Ellen C., 'Tributes to Friedrich Ratzel', *Bulletin of the American Geographical Society* 36/9 (1904), pp. 550–53.

'Geography in the summer schools', *Bulletin of the American Geographical Society* 47/5 (1915), pp. 370–72.

'George Byron Roorbach', *Geographical Review* 24/3 (1934), p. 506.

Gingerich, Owen, *An Annotated Census of Copernicus' De Revolutionibus (Nuremberg, 1543 and Basel, 1566)* (Leiden: Brill, 2002).

———, *The Book Nobody Read: In Pursuit of the Revolutions of Nicolaus Copernicus* (London: William Heinemann, 2004).

'Giuseppe Dalla Vedova', *The Geographical Journal* 55/2 (1920), pp. 157–58.

Glacken, Clarence J., *Traces on the Rhodian Shore: Nature and Culture in Western Thought From Ancient Times to the End of the Eighteenth Century* (Berkeley and Los Angeles: University of California Press, 1967).

Glick, Thomas F. (ed.), *The Comparative Reception of Darwinism* (Austin and London: University of Texas Press, 1974).

Golding, Alan C., 'The Dial, The Little Review, and the dialogics of modernism', *American Periodicals* 15/1 (2005), pp. 42–55.

Goode, J. Paul, 'The human response to the physical environment', *The Journal of Geography* 3/7 (1904), pp. 333–43.

Graff, Gerald, 'Is there a text in this class?', in H. A. Vesser (ed.), *The Stanley Fish Reader* (Malden, MA and Oxford: Blackwell, 1999), pp. 38–54.

Graham, Brian J., 'The search for the common ground: Estyn Evans's Ireland', *Transactions of the Institute of British Geographers* 19/2 (1994), pp. 183–201.

Grossman, Lary, 'Man-environment relationships in anthropology and geography', *Annals of the Association of American Geographers* 67/1 (1977), pp. 126–44.

Guyot, Arnold H., *Earth and Man: Lectures on Comparative Physical Geography in its Relation to the History of Mankind* (Boston: Gould, Kendall and Lincoln, 1849).

Haller, John S., 'The species problem: nineteenth-century concepts of racial inferiority in the origin of man controversy', *American Anthropologist* 72/6 (1970), pp. 1319–29.

Hankins, Thomas L., 'In defence of biography: the use of biography in the history of science', *History of Science* 17/1 (1979), pp. 1–16.

Hardwick, Lorna, *Reception Studies* (Oxford: Oxford University Press, 2003).

Harris, Chauncy D., 'Geographers in the U.S. government in Washington, DC, during World War II', *The Professional Geographer* 49/2 (1997), pp. 245–56.

——, 'Geography at Chicago in the 1930s and 1940s', *Annals of the Association of American Geographers* 69/1 (1979), pp. 21–32.

——, 'Stephen Barr Jones, 1903–1984', *Annals of the Association of American Geographers* 75/2 (1985), pp. 271–76.

Harris, Chauncy D. and Fellmann, Jerome D., 'Geographical serials', *Geographical Review* 40/4 (1950), pp. 649–56.

Harrison, Lowell H., *Western Kentucky University* (Lexington: University Press of Kentucky, 1987).

Harrison, Lowell H. and Klotter, James C., *A New History of Kentucky* (Lexington: University Press of Kentucky, 1997).

Hart, Albert B., 'Review of *Geographic Influences in American History*, by Albert P. Brigham and *American History and its Geographic Conditions*, by Ellen C. Semple', *The American Historical Review* 9/3 (1904), pp. 571–72.

Hartshorne, Richard, *The Nature of Geography: A Critical Survey of Current Thought in the Light of the Past* (Lancaster, PA: Association of American Geographers, 1939).

——, 'Robert S. Platt, 1891–1964', *Annals of the Association of American Geographers* 54/5 (1964), pp. 630–637.

Hawley, Arthur J., 'Environmental perception: nature and Ellen Churchill Semple', *The Southeastern Geographer* 8 (1968), pp. 54–59.

Herbertson, Andrew J., 'Geography in the university', *Scottish Geographical Magazine* 18/3 (1902), pp. 124–32.

——, 'The major natural regions: an essay in systematic geography', *The Geographical Journal* 25/3 (1905), pp. 300–10.

——, 'Two books on the historical geography of the United States. Review of *Geographic Influences in American History*, by Albert P. Brigham and *American History and its Geographic Conditions*, by Ellen C. Semple', *The Geographical Journal* 23/5 (1904), pp. 674–77.

Herman, Theodore, 'George Babcock Cressey, 1896–1963', *Annals of the Association of American Geographers* 55/2 (1965), pp. 360–64.

Hevly, Bruce, 'The heroic science of glacier motion', *Osiris* 11 (1996), pp. 66–86.

Hindle, Wilfrid H., The Morning Post, *1772–1937: Portrait of a Newspaper* (London: G. Routledge & Sons, 1937).

Hooker, Richard, *The Story of an Independent Newspaper* (New York: The Macmillian Company, 1924).

Hubbard, George D., 'Sketch of the life and work of Dr. Frank Carney', *The Ohio Journal of Science* 35/4 (1935), pp. 273–74.

'Human geography. Review of *Influences of Geographic Environment*, by Ellen C. Semple', *Nature* 88/2195 (1911), p. 101.

Hunter, James M., *Perspectives on Ratzel's Political Geography* (Lanham, MD: University Press of America, 1983).

Jackson, Heather J., '"Marginal frivolities": readers' notes as evidence for the history of reading', in R. Myers, M. Harris, and G. Mandelbrote (eds), *Owners, Annotators and the Signs of Reading* (London and New Castle, DE: The British Library and Oak Knoll Press, 2005), pp. 137–51.

Jacobs, Wilbur R., 'Colonial origins of the United States: the Turnerian view', *The Pacific Historical Review* 40/1 (1971), pp. 21–37.

James, Edward T., James, Janet W., and Boyer, Paul S. (eds), *Notable American Women 1607–1950. A Biographical Dictionary*, vol. 3 (Cambridge, MA: Belknap Press, 1971).

James, Preston E., 'Albert Perry Brigham 1855–1932', in T. W. Freeman and P. Pinchemel (eds), *Geographers: Biobibliographical Studies*, vol. 2 (London: Mansell, 1978), pp. 13–17.

——, 'The Blackstone Valley: a study in chorography in Southern New England', *Annals of the Association of American Geographers* 19/2 (1929), pp. 67–109.

——, 'A geographic reconnaissance of Trinidad', *Economic Geography* 3/1 (1927), pp. 87–109.

——, 'Geographical ideas in America, 1890–1914', in B. W. Blouet (ed.), *The Origins of Academic Geography in the United States* (Hamden, CT: Archon Books, 1981), pp. 319–26.

——, 'George Babcock Cressey 1896–1963', *Geographical Review* 54/2 (1964), pp. 254–57.

James, Preston E., Bladen, Wilford A., and Karan, Pradyumna P., 'Ellen Churchill Semple and the development of a research paradigm', in W. A. Bladen and P. P. Karan (eds), *The Evolution of Geographic Thought in America: A Kentucky Root* (Dubuque, IA: Kendall/Hunt, 1983), pp. 28–57.

James, Preston E. and Ehrenberg, Ralph, 'The original members of the Association of American Geographers', *The Professional Geographer* 27/3 (1975), pp. 327–34.

James, Preston E. and Martin, Geoffrey J., *All Possible Worlds: A History of Geographical Ideas*, 2nd ed. (New York: John Wiley & Sons, 1981).

——, *The Association of American Geographers, the First Seventy-Five Years, 1904–1979* (Washington, DC: The Association of American Geographers, 1978).

——, 'On AAG history', *The Professional Geographer* 31/4 (1979), pp. 353–57.

James, Preston E. and Perejda, Andrew D., 'George Babcock Cressey 1896–1963', in T. W. Freeman (ed.), *Geographers: Biobibliographical Studies*, vol. 5 (London: Mansell, 1981), pp. 21–25.

Jeffries, Ella, 'The dependence of the social sciences upon geographical principles', *The Journal of Geography* 26/6 (1926), pp. 228–36.

Johns, Adrian, *The Nature of the Book: Print and Knowledge in the Making* (Chicago and London: University of Chicago Press, 1998).

Johnson, Joan M., *Southern Women at the Seven Sister Colleges: Feminist Values and Social Activism, 1875–1915* (Athens and London: University of Georgia Press, 2008).

Jones, Wellington D. and Sauer, Carl O., 'Outline for field work in geography', *Bulletin of the American Geographical Society* 47/7 (1915), pp. 520–25.

Karan, Pradyumna P. and Mather, Cotton, *Leaders in American Geography*, vol. 2 (Mesilla: The New Mexico Geographical Society, 2000).

Kearns, Gerry, 'Closed space and political practice: Frederick Jackson Turner and Halford Mackinder', *Environment and Planning D: Society and Space* 2/1 (1984), pp. 23–34.

——, 'Halford John Mackinder 1861–1947', in T. W. Freeman (ed.), *Geographers: Biobibliographical Studies*, vol. 9 (London: Mansell, 1985), pp. 71–86.

——, 'The imperial subject: geography and travel in the work of Mary Kingsley and Halford Mackinder', *Transactions of the Institute of British Geographers* 22/4 (1997), pp. 450–72.

Keasby, Lindley M., 'Review of *American History and its Geographic Conditions*, by Ellen C. Semple and *Geographic Influences in American History* by Albert P. Brigham', *Political Science Quarterly* 19/3 (1904), pp. 501–02.

Keighren, Innes M., 'Breakfasting with William Morris Davis: everyday episodes in the history of geography', in E. A. Gagen, H. Lorimer, and A. Vasudevan (eds), *Practising the Archive: Reflections on Methods and Practice in Historical Geography* (London: Royal Geographical Society, 2007), pp. 47–55.

——, 'Bringing geography to the book: charting the reception of *Influences of Geographic Environment*', *Transactions of the Institute of British Geographers* 31/4 (2006), pp. 525–40.

——, 'Giving voice to geography: popular lectures and the diffusion of knowledge', *Scottish Geographical Journal* 124/2 & 3 (2008), pp. 198–203

Keller, Albert G., 'Review of *Influences of Geographic Environment*, by Ellen C. Semple', *The Yale Review* 1 (1912), pp. 331–34.

Keltie, John S., *The Position of Geography in British Universities*, W. L. G. Joerg (ed.) (New York: Oxford University Press, 1921).

——, 'Thirty years' progress in geographical education', *The Geographical Teacher* 7/38 (1914), pp. 215–27.

Keltie, John S. and Howarth, Osbert J. R., *History of Geography* (London: Watts, 1913).

Kenzer, Martin S., 'Milieu and the "intellectual landscape": Carl O. Sauer's undergraduate heritage', *Annals of the Association of American Geographers* 75/2 (1985), pp. 258–70.

Kirby, Harold, 'Charles Atwood Kofoid 1865–1947', *Science* 106/2759 (1947), pp. 462–63.

Kish, George, 'Roberto Almagià', *Geographical Review* 52/4 (1962), pp. 611–12.

Kleber, John E. (ed.), *The Encyclopedia of Louisville* (Lexington: University Press of Kentucky, 2000).

Kniffen, Fred B., 'Richard Joel Russell, 1895–1971', *Annals of the Association of American Geographers* 63/2 (1973), pp. 241–49.

Koelsch, William A., 'Academic geography, American style: an institutional perspective', in G. S. Dunbar (ed.), *Geography: Discipline, Profession and Subject Since 1870* (Dordrecht: Kluwer Academic, 2001), pp. 245–79.

——, *Clark University, 1887–1987: A Narrative History* (Worcester, MA: Clark University Press, 1987).

——, 'Derwent Stainthorpe Whittlesey 1890–1956', in P. H. Armstrong and G. J. Martin (eds), *Geographers: Biobibliographical Studies*, vol. 25 (London: Continuum, 2006), pp. 138–58.

——, 'East and Midwest in American academic geography: two prosopographic notes', *The Professional Geographer* 53/1 (2001), pp. 97–105.

——, 'Franz Boas, geographer, and the problem of disciplinary identity', *Journal of the History of the Behavioural Sciences* 40/1 (2004), pp. 1–22.

——, 'Geography at Clark: the first fifty years, 1921–1971', in J. E. Harmon and T. J. Richards (eds), *Geography in New England* (New Britain, CT: New England/St Lawrence Valley Geographical Society, 1988).

——, 'The historical geography of Harlan H. Barrows', *Annals of the Association of American Geographers* 59/4 (1969), pp. 632–51.

——, 'Seedbed of reform: Arnold Guyot and school geography in Massachusetts, 1849–1855', *The Journal of Geography* 107/2 (2008), pp. 35–42.

——, 'Three friends of Swiss-American science: Louis Agassiz, Arnold Guyot, and Cornelius C. Felton', *Swiss American Historical Society Review* 44/1 (2008), pp. 45–59.

——, 'Wallace Atwood's "Great Geographical Institute"', *Annals of the Association of American Geographers* 70/4 (1980), pp. 567–82.

Kofoid, Charles A., 'Environment as a force in human action', *The Dial* 51/610 (1911), pp. 398–99.

Kohlstedt, Sally G., 'Concepts of place in museum space at the Smithsonian Institution's National Museum' (paper presented at the Geographies of Nineteenth-Century Science, University of Edinburgh, 2007).

——, '"Thoughts in things": modernity, history, and North American museums', *Isis* 96/4 (2005), pp. 586–601.

Kolb, Albert, 'Ferdinand Freiherr von Richthofen 1833–1905', in T. W. Freeman (ed.), *Geographers: Biobibliographical Studies*, vol. 7 (London: Mansell, 1983), pp. 109–15.

Krauss, Rosalind, 'Photography's discursive spaces', in J. Evans and S. Hall (eds), *Visual Culture: The Reader* (London: Sage, 1989), pp. 193–209.

Kroll, Barry M., 'Writing for readers: three perspectives on audience', *College Composition and Communication* 35/2 (1984), pp. 172–85.

Leighly, John, 'Carl Ortwin Sauer 1889–1975', in T. W. Freeman and P. Pinchemel (eds), *Geographers: Biobibliographical Studies*, vol. 2 (London: Mansell, 1978), pp. 99–108.

——, 'Carl Ortwin Sauer, 1889–1975', *Annals of the Association of American Geographers* 66/3 (1976), pp. 337–48.

——, 'Drifting into geography in the Twenties', *Annals of the Association of American Geographers* 69/1 (1979), pp. 4–9.

——, 'Review of *The Geographic Factor: Its Rôle in Life and Civilization*, by Ray H. Whitbeck and Olive J. Thomas', *American Anthropologist* 35/4 (1933), pp. 766–67.

——, 'What has happened to physical geography?', *Annals of the Association of American Geographers* 45/4 (1955), pp. 309–18.

——, (ed.), *Land and Life: A Selection of Writings of Carl Ortwin Sauer* (Berkeley: University of California Press, 1963).

Lewis, Carolyn B., 'The biography of a neglected classic: Ellen Churchill Semple's *The Geography of the Mediterranean Region*' (Ph.D. diss., University of South Carolina, 1979).

Libby, Orin G., 'Review of *Influences of Geographic Environment*, by Ellen C. Semple', *The American Historical Review* 17/2 (1912), pp. 355–57.

Lightman, Bernard, 'Lecturing in the spatial economy of science', in A. Fyfe and B. Lightman (eds), *Science in the Marketplace: Nineteenth-Century Sites and Experiences* (Chicago and London: University of Chicago Press, 2007), pp. 97–132.

Linehan, Thomas, *British Fascism, 1918–39: Parties, Ideology and Culture* (Manchester: Manchester University Press, 2000).

Livingstone, David N., *Adam's Ancestors: Race, Religion, and the Politics of Human Origins* (Baltimore: Johns Hopkins University Press, 2008).

——, 'Environment and inheritance: Nathaniel Southgate Shaler and the American frontier', in B. W. Blouet (ed.), *The Origins of Academic Geography in the United States* (Hamden, CT: Archon Books, 1981), pp. 123–38.

——, 'Geographical inquiry, rational religion, and moral philosophy: enlightenment discourses on the human condition', in D. N. Livingstone and C. W. J. Withers (eds), *Geography and Enlightenment* (Chicago and London: University of Chicago Press, 1999), pp. 93–119.

——, *The Geographical Tradition: Episodes in the History of a Contested Enterprise* (Oxford: Blackwell, 1992).

——, 'The history of science and the history of geography: interactions and implications', *History of Science* 22/3 (1984), pp. 271–302.

——, 'Mobilising science: writers, readers and the geographies of meaning' (paper presented at the Geographies of Nineteenth-Century Science, University of Edinburgh, 2007).

——, *Nathaniel Southgate Shaler and the Culture of American Science* (Tuscaloosa: University of Alabama Press, 1987).

——, *Putting Science in Its Place: Geographies of Scientific Knowledge* (Chicago and London: University of Chicago Press, 2003).

——, '"Risen into empire": moral geographies of the American Republic', in D. N. Livingstone and C. W. J. Withers (eds), *Geography and Revolution* (Chicago: University of Chicago Press, 2005).

——, 'Science, religion and the geography of reading: Sir William Whitla and the editorial staging of Isaac Newton's writings on biblical prophecy', *The British Journal for the History of Science* 36/1 (2003), pp. 27–42.

——, 'Science, site and speech: scientific knowledge and the spaces of rhetoric', *History of the Human Sciences* 20/2 (2007), pp. 71–98.

——, 'Science, text and space: thoughts on the geography of reading', *Transactions of the Institute of British Geographers* 30/4 (2005), pp. 391–401.

——, 'Text, talk and testimony: geographical reflections on scientific habits. An afterword', *British Journal for the History of Science* 38/1 (2005), pp. 93–100.

Lloyd, Genevieve, *The Man of Reason: 'Male' and 'Female' in Western Philosophy* (London: Methuen, 1984).

Lowenthal, David, 'Fruitful liaison or folie à deux? The AAG and AGS', *The Professional Geographer* 57/3 (2005), pp. 468–73.

——, *George Perkins Marsh: Prophet of Conservation* (Seattle: University of Washington Press, 2000).

Lukacs, John, *Destinations Past* (Columbia: University of Missouri Press, 1994).

Luxenberg, Adele, 'Women at Leipzig', *The Nation*, 4 October 1894, pp. 247–48.

Luzzana Caraci, Ilaria, 'Modern geography in Italy: from the archives to environmental management', in G. S. Dunbar (ed.), *Geography: Discipline, Profession and Subject Since 1870* (Dordrecht: Kluwer, 2001), pp. 121–51.

Lyell, Charles, *Travels in North America; With Geological Observations on the United States, Canada, and Nova Scotia*, vol. 1 (London: John Murray, 1845).

MacDonald, Bertrum H. and Black, Fiona A., 'Using GIS for spatial and temporal analyses in print culture studies', *Social Science History* 24/3 (2000), pp. 505–36.

Mackinder, Halford J., 'Modern geography, German and English', *The Geographical Journal* 6/4 (1895), pp. 367–79.

MacLean, Kenneth, 'George Goudie Chisholm 1850–1930', in T. W. Freeman (ed.), *Geographers: Biobibliographical Studies*, vol. 12 (London: Mansell, 1988), pp. 21–33.

Macpherson, Anne, 'Clarence James Glacken 1909–1989', in G. J. Martin (ed.), *Geographers: Biobibliographical Studies*, vol. 14 (London: Mansell, 1993), pp. 27–42.

Maddrell, Avril M. C., *Complex Locations: Women's Geographical Work in the UK 1850–1970* (Chichester: Wiley-Blackwell, 2009).

——, 'Scientific discourse and the geographical work of Marion Newbigin', *Scottish Geographical Magazine* 113/1 (1997), pp. 33–41.

Mahard, Richard H., 'A history of the department of geology and geography, Denison University, Granville, Ohio', *The Ohio Journal of Science* 79/1 (1979), pp. 18–21.

Marsh, George P., *Man and Nature; or, Physical Geography as Modified by Human Action* (New York: Charles Scribner, 1864).

Martin, A. F., 'The necessity for determinism: a metaphysical problem confronting geographers', *Transactions and Papers (Institute of British Geographers)* 17 (1951), pp. 1–11.

Martin, Geoffrey J., *All Possible Worlds: A History of Geographical Ideas*, 4th ed. (Oxford: Oxford University Press, 2005).

——, *Ellsworth Huntington: His Life and Thought* (Hamden, CT: Archon Books, 1973).

——, 'The emergence and development of geographic thought in New England', *Economic Geography* 74 (1998), pp. 1–13.

——, 'Geography, geographers and Yale University, c. 1770–1970', in J. E. Harmon and T. J. Richards (eds), *Geography in New England* (New Britain, CT: New England/St. Lawrence Valley Geographical Society, 1988), pp. 2–9.

——, 'J. Russell Smith 1874–1966', in P. H. Armstrong and G. J. Martin (eds), *Geographers: Biobibliographical Studies*, vol. 21 (London: Mansell, 2001), pp. 97–113.

——, *The Life and Thought of Isaiah Bowman* (Hamden, CT: Archon Books, 1980).

——, 'On Whittlesey, Bowman and Harvard', *Annals of the Association of American Geographers* 78/1 (1988), pp. 152–58.

——, 'Preston E. James, 1899–1986', *Annals of the Association of American Geographers* 78/1 (1988), pp. 164–75.

——, 'Preston Everett James 1899–1986', in T. W. Freeman (ed.), *Geographers: Biobibliographical Studies*, vol. 11 (London: Mansell, 1987), pp. 63–70.

Mason, Otis T., 'Resemblances in arts widely separated', *The American Naturalist* 20/3 (1886), pp. 246–51.

Mather, Kirtley F., 'Evolution and religion', *The Scientific Monthly* 21/3 (1925), pp. 322–28.

Mathewson, Kent and Shoemaker, Vincent J., 'Louisiana State University geography at seventy-five: "Berkeley on the bayou" and beyond', in J. O. Wheeler and S. D. Brunn (eds), *The Role of the South in the Making of American Geography: Centennial of the AAG, 2004* (Columbia, MD: Bellweather, 2004), pp. 245–67.

——, 'Louisiana State University's Department of Geography and Anthropology: a selective history', *Southwestern Geographer* 1 (1997), pp. 62–84.

Mayhew, Robert J., 'The character of English geography c. 1660–1800: a textual approach', *Journal of Historical Geography* 24/4 (1998), pp. 385–412.

——, 'Denaturalising print, historicising text: historical geography and the history of the book', in E. A. Gagen, H. Lorimer, and A. Vasudevan (eds), *Practising the Archive: Reflections on Methods and Practice in Historical Geography* (London: Royal Geographical Society, 2007), pp. 23–36.

——, 'Mapping science's imagined community: geography as a Republic of Letters, 1600–1800', *British Journal for the History of Science* 38/1 (2005), pp. 73–92.

——, 'Materialist hermeneutics, textuality and the history of geography: print spaces in British geography, c. 1500–1900', *Journal of Historical Geography* 33/3 (2007), pp. 466–88.

McEwan, Cheryl, 'Cutting power lines within the palace? Countering paternity and eurocentrism in the "geographical tradition"', *Transactions of the Institute of British Geographers* 23/3 (1998), pp. 371–84.

——, 'Gender, science and physical geography in nineteenth-century Britain', *Area* 30/3 (1998), pp. 215–23.

McIntire, William G., 'Richard Joel Russell (1895–1971)', *Geographical Review* 63/2 (1973), pp. 276–79.

McKee, Jesse O., 'Interview with Fred B. Kniffen', *The Mississippi Geographer* 4/1 (1976), pp. 5–6.

McManis, Douglas R., 'Leading ladies at the AGS', *Geographical Review* 86/2 (1996), pp. 270–77.

Meadows, Paul, 'Achille Loria: agrarian determinist', *American Journal of Economics and Sociology* 10/2 (1951), pp. 175–54.

'Meeting in Kentucky', *The Journal of Geography* 15/10 (1917), p. 336.

'The meeting of the American Historical Association at Madison', *The American Historical Review* 13/3 (1908), pp. 433–58.

'Meeting of the Society', *Bulletin of the American Geographical Society* 45/4 (1913), p. 285.

Meller, Helen, 'Geddes, Sir Patrick (1854–1932)', in H. C. G. Matthew and B. Harrison (eds), *Oxford Dictionary of National Biography* (Oxford: Oxford University Press, 2004).

——, *Patrick Geddes: Social Evolutionist and City Planner* (London: Routledge, 1993).

Memorial Volume of the Transcontinental Excursion of 1912 of the American Geographical Society of New York (New York: American Geographical Society, 1915).

Merrill, James A., 'A suggestive course in geography', *Journal of School Geography* 2/9 (1898), pp. 321–28.

Mikesell, Marvin W., 'Continuity and change', in B. W. Blouet (ed.), *The Origins of Academic Geography in the United States* (Hamden, CT: Archon Books, 1981), pp. 1–15.

——, 'Geographic perspectives in anthropology', *Annals of the Association of American Geographers* 57/3 (1967), pp. 617–34.

——, 'Ratzel, Friedrich', in D. L. Sills (ed.), *International Encyclopedia of Social Sciences*, vol. 13 (New York: MacMillan and Free Press, 1968), pp. 327–29.

Millett, Frederick B., *Contemporary American Authors: A Critical Survey and 219 Bio-Bibliographies* (London: Harrap, 1940).

Monk, Janice J., 'The women were always welcome at Clark', *Economic Geography* 74 (1998), pp. 14–30.

——, 'Women, gender, and the histories of American geography', *Annals of the Association of American Geographers* 94/1 (2004), pp. 1–22.

——, 'Women's worlds at the American Geographical Society', *Geographical Review* 93/2 (2004), pp. 237–57.

Monk, Janice J., Fortuijn, Joos D., and Raleigh, Clionadh, 'The representation of women in academic geography: contexts, climate and curricula', *Journal of Geography in Higher Education* 28/1 (2004), pp. 83–90.

Murphy, David T., '"A sum of the most wonderful things": Raum, geopolitics and the German tradition of environmental determinism, 1900–1933', *History of European Ideas* 25/3 (1999), pp. 121–33.

Nash, Peter H., Guelke, Leonard, and Preston, Richard E., *Abstract Thoughts, Concrete Solutions: Essays in Honour of Peter Nash* (Waterloo, Ontario: Department of Geography, University of Waterloo, 1987).

Natter, Wolfgang, 'Friedrich Ratzel's spatial turn: identities of disciplinary space and its borders between the anthropo- and political geography of Germany and the United States', in H. van Houtum, O. Kramsch, and Z. Wolfgang (eds), *Bordering Space* (Aldershot: Ashgate, 2005), pp. 171–88.

Naylor, Simon, 'The field, the museum and the lecture hall: the spaces of natural history in Victorian Cornwall', *Transactions of the Institute of British Geographers* 27/4 (2002), pp. 494–513.

Negri, Paul (ed.), *Great Short Poems* (Mineola, NY: Dover, 2000).

Nelson, Howard J., 'J. E. Spencer, 1907–1984', *Annals of the Association of American Geographers* 75/4 (1985), pp. 595–603.

Nelson, Jack, 'New Orleans *Times-Picayune* series on racism', in T. Rosensteil and A. S. Mitchell (eds), *Thinking Clearly: Cases in Journalistic Decision-Making* (New York: Columbia University Press, 2003).

Nelson, Robert S., 'The slide lecture, or the work of art "history" in the age of mechanical reproduction', *Critical Inquiry* 26/3 (2000), pp. 414–34.

Nersessian, Nancy J., 'Opening the black box: cognitive science and history of science', *Osiris* 10 (1995), pp. 194–211.

Numbers, Ronald L. and Stephens, Lester D., 'Darwinism in the American south', in R. L. Numbers and J. Stenhouse (eds), *Disseminating Darwinism: The Role of Place, Race, Religion, and Gender* (Cambridge: Cambridge University Press, 1999), pp. 123–43.

Ogborn, Miles, '*Geographia*'s pen: writing, geography and the arts of commerce, 1660–1760', *Journal of Historical Geography* 30/2 (2004), pp. 294–315.

——, 'Writing travels: power, knowledge and ritual on the English East India Company's early voyages', *Transactions of the Institute of British Geographers* 27/2 (2002), pp. 155–71.

Ogborn, Miles and Withers, Charles W. J., 'Travel, trade, and empire: knowing other places, 1660–1800', in C. Wall (ed.), *A Concise Companion to the Restoration and Eighteenth Century* (Oxford: Blackwell, 2005), pp. 13–35.

——, (eds), *Geographies of the Book* (Farnham: Ashgate, 2010).

Palmieri, Patricia A., 'Here was fellowship: a social portrait of academic women at Wellesley College, 1895–1920', *History of Education Quarterly* 32/2 (1983), pp. 195–214.

Pattison, William D., 'Goode's proposal of 1902: an interpretation', *The Professional Geographer* 30/1 (1978), pp. 3–8.

——, 'Rollin D. Salisbury 1858–1922', in T. W. Freeman (ed.), *Geographers: Bio-bibliographical Studies*, vol. 6 (London: Mansell, 1982).

——, 'Rollin Salisbury and the establishment of geography at the University of Chicago', in B. W. Blouet (ed.), *The Origins of Academic Geography in the United States* (Hamden, CT: Archon Books, 1981), pp. 151–63.

Paty, Michel, 'The scientific reception of relativity in France', in T. F. Glick (ed.), *The Comparative Reception of Relativity* (Dordrecht: D. Reidel, 1987), pp. 113–67.

Peet, Richard, 'The social origins of environmental determinism', *Annals of the Association of American Geographers* 75/3 (1985), pp. 309–33.

Platt, Robert S., 'Wellington Downing Jones', *Geographical Review* 48/2 (1958), pp. 285–87.

'Presentation of the Cullum Geographical Medal to Hugh Robert Mill', *Geographical Review* 20/4 (1930), pp. 669–70.

Price, Edward T., 'Geography at the University of Oregon', *Association of Pacific Coast Geographers Yearbook* 52 (1990), pp. 140–52.

Price, Leah, 'Review of *In Another Country: Colonialism, Culture, and the English Novel in India*, by Priya Joshi', *Victorian Studies* 45/2 (2003), pp. 333–34.

Prunty, Merle C., 'Clark in the early 1940s', *Annals of the Association of American Geographers* 69/1 (1979), pp. 42–45.

Radcliffe, Sarah A., Watson, Elizabeth E., Simmons, Ian, Fernández-Armesto, Felipe, and Sluyter, Andrew, 'Forum: Environmentalist thinking and/in geography', *Progress in Human Geography* 34/1 (2009), pp. 98–116.

Ratzel, Friedrich, *Sein und Werden der organischen Welt. Eine populäre Schöpfungsgeschichte* (Leipzig: Gebhart und Reisland, 1869).

——, 'Studies in political areas. The political territory in relation to earth and continent', *The American Journal of Sociology* 3/3 (1897), pp. 297–313.

——, 'Studies in political areas. II. Intellectual, political, and economic effects of large areas', *The American Journal of Sociology* 3/4 (1898), pp. 449–63.

——, 'Studies in political areas. III. The small political area', *The American Journal of Sociology* 4/3 (1898), pp. 366–79.

Raup, Hugh M., 'Trends in the development of geographic botany', *Annals of the Association of American Geographers* 32/4 (1942), pp. 319–54.

Ravenstein, E. G., 'Obituary: Ferdinand Freiherr von Richthofen', *The Geographical Journal* 26/6 (1905), pp. 679–82.

Renwick, Chris, 'The practice of Spencerian science: Patrick Geddes's biosocial program, 1876–1889', *Isis* 100/1 (2009), pp. 36–57.

'Review of *Influences of Geographic Environment*, by Ellen C. Semple', *The Journal of Geography* 10/1 (1911), pp. 33–34.

'Review of *Influences of Geographic Environment*, by Ellen C. Semple', *Scottish Geographical Magazine* 27 (1912), pp. 160–62.

Rich, John L., 'Cultural features and the physiographic cycle', *Geographical Review* 4/4 (1917), pp. 297–308.

Ríos-Font, Wadda C., *The Canon and the Archive: Configuring Literature in Modern Spain* (Lewisburg, PA: Bucknell University Press, 2004).

Ritter, Carl, *Comparative Geography*, tr. W. L. Gage (Philadelphia: J. B. Lippincott & Co., 1865).

'Robert Henry Kinvig', *Transactions of the Institute of British Geographers* 49 (1970), pp. 519–24.

Robinson, Edward van Dyke, 'Review of *Influences of Geographic Environment*, by Ellen C. Semple', *The American Economic Review* 2/2 (1912), pp. 338–40.

Roorbach, George B., 'Review of *Influences of Geographic Environment*, by Ellen C. Semple', *Annals of the American Academy of Political and Social Science* 41 (1912), pp. 350–51.

——, 'The trend of modern geography', *Bulletin of the American Geographical Society* 46/11 (1914), pp. 801–16.

Roosevelt, Theodore, *History as Literature and Other Essays* (New York: Charles Scribner's Sons, 1913).

——, *The Winning of the West*, 4 vols. (New York: G. P. Putnam's, 1889–1896).

Rose, John K., 'Stephen Sargent Visher, 1887–1967', *Annals of the Association of American Geographers* 61/2 (1971), pp. 394–406

Rowley, Virginia M., *J. Russell Smith: Geographer, Educator, and Conservationist* (Philadelphia: University of Pennsylvania Press, 1964).

Rubin, Barbara, '"Women in geography" revisited: present status, new options', *The Professional Geographer* 31/2 (1979), pp. 125–34.

Rupke, Nicolaas A., 'A geography of enlightenment: the critical reception of Alexander von Humboldt's Mexico work', in D. N. Livingstone and C. W.

J. Withers (eds), *Geography and Enlightenment* (Chicago and London: University of Chicago Press, 1999), pp. 319–43.

——, 'Translation studies in the history of science: the example of *Vestiges*', *The British Journal for the History of Science* 33/2 (2000), pp. 209–22.

Russell, Colin A., 'The reception of Newtonianism in Europe', in D. Goodman and C. A. Russell (eds), *The Rise of Scientific Europe 1500–1800* (Sevenoaks: Hodder & Stoughton, 1991), pp. 253–78.

Ryan, James R., 'History and philosophy of geography: bringing geography to book, 2000–2001', *Progress in Human Geography* 27/2 (2003), pp. 195–202.

——, *Picturing Empire: Photography and the Visualization of the British Empire* (London: Reaktion Books, 1997).

Sachs, Aaron J., *The Humboldt Current: Nineteenth-Century Exploration and the Roots of American Environmentalism* (New York: Viking, 2006).

Said, Edward W., *The World, the Text, and the Critic* (London: Vintage, 1991).

Sanguin, André-Louis, 'En relisant Ratzel', *Annales de Géographie* 555 (1990), pp. 579–94.

Sauer, Carl O., 'The education of a geographer', *Annals of the Association of American Geographers* 46/3 (1956), pp. 287–99.

——, 'The formative years of Ratzel in the United States', *Annals of the Association of American Geographers* 61/2 (1971), pp. 245–54.

——, 'The fourth dimension of geography', *Annals of the Association of American Geographers* 64/2 (1974), pp. 189–92.

——, 'Geography of the Upper Illinois Valley and history of development', *Illinois State Geological Survey* Bulletin no. 27 (1916).

——, 'Memorial of Ruliff S. Holway', *Annals of the Association of American Geographers* 19/1 (1929), pp. 64–65.

——, 'The morphology of landscape', *University of California Publications in Geography* 2/2 (1925), pp. 19–54.

——, 'Recent developments in cultural geography', in E. C. Hayes (ed.), *Recent Developments in the Social Sciences* (Philadelphia: J. B. Lippincott, 1927), pp. 154–212.

——, *Seventeenth Century North America* (Berkeley: Turtle Island Foundation, 1980).

——, 'Status and change in the rural Midwest – a retrospect', in W. A. Bladen and P. P. Karan (eds), *The Evolution of Geographic Thought in America: A Kentucky Root* (Dubuque, IA: Kendall/Hunt, 1983), pp. 115–22.

——, 'The survey method in geography and its objectives', *Annals of the Association of American Geographers* 14/1 (1924), pp. 17–33.

Scargill, David I., 'The RGS and the foundations of geography at Oxford', *The Geographical Journal* 142/3 (1976), pp. 438–61.

Schelhaas, Bruno and Hönsch, Ingrid, 'History of German geography: worldwide reputation and strategies of nationalism and institutionalisation', in G. S. Dunbar (ed.), *Geography: Discipline, Profession and Subject Since 1870* (Dordrecht: Kluwer, 2001), pp. 9–44.

Schlüter, Otto, 'Review of *Influences of Geographic Environment*, by Ellen C. Semple', *Petermanns Geographische Mitteilungen* 58 (1912), p. 166.

Schneider, Allan F., 'Chamberlin, Salisbury, and Collie: a tale of three Beloit College geologists', *Geoscience Wisconsin* 18 (2001), pp. 9–20.

Schwartz, Joan M., 'The Geography Lesson: photographs and the construction of imaginative geographies', *Journal of Historical Geography* 22/1 (1996), pp. 16–45.

Scott, Donald M., 'The popular lecture and the creation of a public in mid-nineteenth-century America', *The Journal of American History* 66/4 (1980), pp. 791–809.

Searle, Geoffrey R. (ed.), *Eugenics and Politics in Britain: 1900–1914* (Leyden: Noordhoff International Publishing, 1976).

Secord, James A., 'How scientific conversation became shop talk', *Transactions of the Royal Historical Society* 17 (2007), pp. 129–56.

——, 'Knowledge in transit', *Isis* 95/4 (2004), pp. 654–72.

——, *Victorian Sensation: The Extraordinary Publication, Reception, and Secret Authorship of* Vestiges of the Natural History of Creation (Chicago and London: University of Chicago Press, 2000).

Semple, Ellen C., *American History and its Geographic Conditions* (Boston: Houghton, Mifflin and Company, 1903).

——, 'The Anglo-Saxons of the Kentucky Mountains: a study in anthropogeography', *The Geographical Journal* 17/6 (1901), pp. 561–94.

——, 'The Anglo-Saxons of the Kentucky Mountains: a study in anthropogeography', *Bulletin of the American Geographical Society* 42/8 (1910), pp. 588–623.

——, 'The barrier boundary of the Mediterranean basin and its northern breaches as factors in history', *Annals of the Association of American Geographers* 5 (1915), pp. 27–59.

——, 'Coast peoples. Part I', *The Geographical Journal* 31/1 (1908), pp. 72–90.

——, 'Coast peoples. Part II', *The Geographical Journal* 31/2 (1908), pp. 170–87.

——, 'A comparative study of the Atlantic and Pacific oceans. Part I', *Journal of School Geography* 3/4 (1899), pp. 121–29.

——, 'A comparative study of the Atlantic and Pacific oceans. Part II', *Journal of School Geography* 3/5 (1899), pp. 172–80.

——, 'The development of Hanse towns in relation to their geographical environment', *Journal of the American Geographical Society of New York* 31/3 (1899), pp. 236–55.

——, 'Domestic and municipal waterworks in ancient Mediterranean lands', *Geographical Review* 21/3 (1931), pp. 466–74.

——, 'Emphasis upon anthropo-geography in schools', *The Journal of Geography* 3/8 (1904), pp. 366–74.

——, 'Geographical boundaries. Part I', *Bulletin of the American Geographical Society* 39/7 (1907), pp. 385–97.

——, 'Geographical boundaries. Part II', *Bulletin of the American Geographical Society* 39/8 (1907), pp. 449–63.

——, 'Geographical location as a factor in history', *Bulletin of the American Geographical Society* 40/2 (1908), pp. 65–81.

——, 'The Indians of southeastern Alaska in relation to their environment', *Journal of School Geography* 2/6 (1898), pp. 206–15.

——, 'The influence of geographic conditions upon ancient Mediterranean stock-raising', *Annals of the Association of American Geographers* 13 (1922), pp. 3–38.

——, 'The influence of the Appalachian barrier upon colonial history', *Journal of School Geography* 1 (1897), pp. 33–41.

——, *Influences of Geographic Environment on the Basis of Ratzel's System of Anthropo-geography* (New York: Henry Holt and Company, 1911).

——, 'Japanese colonial methods', *Bulletin of the American Geographical Society* 45/4 (1913), pp. 255–75.

——, 'Louisville, a study in economic geography', *Journal of School Geography* 4 (1900), pp. 361–70.

——, 'A new departure in social settlements', *Annals of the American Academy of Political and Social Science* 15 (1900), pp. 157–60.

——, 'Oceans and enclosed seas: a study in anthropo-geography', *Bulletin of the American Geographical Society* 40/4 (1908), pp. 193–209.

——, 'The operation of geographic factors in history', *Bulletin of the American Geographical Society* 41/7 (1909), pp. 422–39.

——, 'Pirate coasts of the Mediterranean Sea', *Geographical Review* 2/2 (1916), pp. 134–51.

——, 'Review of *Anthropogeographie* by Friedrich Ratzel', *Annals of the American Academy of Political and Social Science* 16 (1900), pp. 137–39.

——, 'Review of *Die staat und sein boden*, by Friedrich Ratzel', *Annals of the American Academy of Political and Social Science* 9 (1897), pp. 102–04.

——, 'Review of *Politische geographie der vereinigten staaten von Amerika*, by Friedrich Ratzel', *Annals of the American Academy of Political and Social Science* 4 (1894), pp. 139–40.

——, 'Some geographic causes determining the location of cities', *Journal of School Geography* 2 (1897), pp. 206–31.

Shaler, Nathaniel S., *Nature and Man in America* (New York: Charles Scribner's Sons, 1891).

Shapin, Steven, 'Man with a plan: Herbert Spencer's theory of everything', *The New Yorker*, 13 August 2007, pp. 75–79.

Shapiro, Henry D., *Appalachia on Our Mind: The Southern Mountains and Mountaineers in the American Consciousness, 1870–1920* (Chapel Hill: University of North Carolina Press, 1986).

Shrock, Robert R., *Geology at M.I.T., 1865–1965: A History of the First Hundred Years of Geology at Massachusetts Institute of Technology* (Cambridge, MA: The MIT Press, 1982).

Simon, Walter M., 'Herbert Spencer and the "social organism"', *Journal of the History of Ideas* 21/2 (1960), pp. 294–99.

Simpson, Robert E. and Huke, Robert E., 'One view of geography at Dartmouth College', in J. E. Harmon and T. J. Richards (eds), *Geography in New England* (New Britain, CT: New England/St Lawrence Valley Geographical Society, 1988), pp. 10–24.

Sluyter, Andrew, 'Neo-environmental determinism, intellectual damage control, and nature/society science', *Antipode* 35/4 (2003), pp. 813–17.

Smith, Ben A. and Vining, James W., *American Geographers, 1784–1812: A Bio-Bibliographical Guide* (Westport, CT: Praeger Publishers, 2003).

Smith, J. Russell, 'Economic geography and its relation to economic theory and higher education', *Bulletin of the American Geographical Society* 39/8 (1907), pp. 472–81.

Smith, Neil, '"Academic war over the field of geography": the elimination of geography at Harvard, 1947–1951', *Annals of the Association of American Geographers* 77/2 (1987), pp. 155–72.

——, *American Empire: Roosevelt's Geographer and the Prelude to Globalization* (Berkeley: University of California Press, 2003).

——, 'Geography as museum: private history and conservative idealism in *The Nature of Geography*', in J. N. Entrikin and S. D. Brunn (eds), *Reflections on Richard Hartshorne's* The Nature of Geography (Washington: Association of American Geographers, 1989), pp. 91–120.

Smith, Woodruff D., 'Friedrich Ratzel and the origins of Lebensraum', *German Studies Review* 3/1 (1980), pp. 51–68.

Solot, Michael, 'Carl Sauer and cultural evolution', *Annals of the Association of American Geographers* 76/4 (1986), pp. 508–20.

Spencer, Joseph E., 'A geographer west of the Sierra Nevada', *Annals of the Association of American Geographers* 69/1 (1979), pp. 46–52.

Speth, William W., 'Berkeley geography, 1923–33', in B. W. Blouet (ed.), *The Origins of Academic Geography in the United States* (Hamden, CT: Archon Books, 1981), pp. 221–44.

——, *How it Came to Be: Carl O. Sauer, Franz Boas and the Meanings of Anthropogeography* (Ellensburg: Ephemera Press, 1999).

Stadtman, Verne A. (ed.), *The Centennial Record of the University of California* (Berkeley, CA: University of California Printing Department, 1967).

Steel, Robert W., 'John Norman Leonard Baker 1893–1971', in G. J. Martin (ed.), *Geographers: Biobibliographical Studies*, vol. 16 (London: Mansell, 1995).

Steward, Julian H., Gibson, Ann J., and Rowe, John H., 'Alfred Louis Kroeber, 1876–1960', *American Anthropologist* 63/5 (1961), pp. 1038–87.

Stoddart, David R., 'Darwin's impact on geography', *Annals of the Association of American Geographers* 56/4 (1966), pp. 683–98.

'Summer session courses in geography', *Geographical Review* 3/6 (1917), pp. 491–96.

Szefel, Lisa, 'Encouraging verse: William S. Braithwaite and the poetics of race', *The New England Quarterly* 74/1 (2001), pp. 32–61.

Taft, Donald R., 'Alvan A. Tenney: a tribute', *American Sociological Review* 2/1 (1937), pp. 67–68.

Tang, Chenxi, *The Geographical Imagination of Modernity: Geography, Literature, and Philosophy in German Romanticism* (Palo Alto, CA: Stanford University Press, 2008).

Tarr, Ralph S., 'Review of *American History and its Geographic Conditions*, by Ellen C. Semple', *Bulletin of the American Geographical Society* 35/5 (1903), pp. 566–70.

Tatham, George, 'Environmentalism and possibilism', in G. Taylor (ed.), *Geography in the Twentieth Century* (London: Methuen, 1957), pp. 128–62.

——, 'Geography in the nineteenth century', in G. Taylor (ed.), *Geography in the Twentieth Century* (London: Methuen, 1957), pp. 28–69.

Taylor, Eva G. R., 'Review of *The Geographic Background of Greek and Roman History*, by Max Cary', *The Geographical Journal* 114/1–3 (1949), pp. 83–84.

Tenney, Alvan A., 'Review of *Influences of Geographic Environment*, by Ellen C. Semple', *Political Science Quarterly* 27/2 (1912), pp. 345–48.

Thoman, Richard S., 'Robert Swanton Platt 1891–1964', in T. W. Freeman and P. Pinchemel (eds), *Geographers: Biobibliographical Studies*, vol. 3 (London: Mansell, 1979), pp. 107–16.

Thompson, Lawrence S., 'Alice Caldwell Hegan Rice', in R. Bain, J. M. Flora, and L. D. Rubin (eds), *Southern Writers: A Biographical Dictionary* (1979), pp. 381–82.

Topham, Jonathan R., 'Scientific publishing and the reading of science in nineteenth-century Britain: a historiographical survey and guide to sources', *Studies in History and Philosophy of Science* 31/4 (2000), pp. 559–612.

——, 'A view from the industrial age', *Isis* 95/3 (2004), pp. 431–42.

Trewartha, Glenn T., 'Geography at Wisconsin', *Annals of the Association of American Geographers* 69/1 (1979), pp. 16–21.

Trindell, Roger T., 'Franz Boas and American geography', *The Professional Geographer* 21/5 (1969), pp. 328–32.

Tucker, Jennifer, 'Photography as witness, detective, and impostor: visual representation in Victorian science', in Bernard Lightman (ed.), *Victorian Science in Context* (Chicago and London: University of Chicago Press, 1997), pp. 378–408.

Turner, Frederick J., 'Geographical interpretations of American history. Review of *American History and its Geographic Conditions*, by Ellen C. Semple and *Geographic Influences in American History*, by Albert P. Brigham', *The Journal of Geography* 4/1 (1905), pp. 34–37.

van Wyhe, John, 'The diffusion of phrenology through public lecturing', in A. Fyfe and B. Lightman (eds), *Science in the Marketplace: Nineteenth-Century Sites and Experiences* (Chicago and London: University of Chicago Press, 2007), pp. 60–96.

Vandenberg, Peter, 'Coming to terms: audience', *The English Journal* 84/4 (1995), pp. 79–80.

Visher, Stephen S., 'Climatic influences', in G. Taylor (ed.), *Geography in the Twentieth Century* (London: Methuen, 1957), pp. 196–220.

——, 'Richard Elwood Dodge, 1868–1952', *Annals of the Association of American Geographers* 42/4 (1952), pp. 318–21.

——, 'Rollin D. Salisbury and geography', *Annals of the Association of American Geographers* 43/1 (1953), pp. 4–11.

Walker, H. Jesse, 'Evelyn Lord Pruitt, 1918–2000', *Annals of the Association of American Geographers* 96/2 (2006), pp. 432–39.

——, 'Richard Joel Russell 1895–1971', in T. W. Freeman and P. Pinchemel (eds), *Geographers: Biobibliographical Studies*, vol. 4 (London: Mansell, 1980), pp. 127–38.

Walker, H. Jesse and Richardson, Miles E., 'Fred Bowerman Kniffen, 1900–1993', *Annals of the Association of American Geographers* 84/4 (1994), pp. 732–43.

Waller, Philip, *Writers, Readers, and Reputations: Literary Life in Britain 1870–1918* (Oxford: Oxford University Press, 2006).

Walls, Laura D., *The Passage to Cosmos: Alexander von Humboldt and the Shaping of America* (Chicago and London: University of Chicago Press, 2009).

Wanklyn, Harriet, *Friedrich Ratzel: A Biographical Memoir and Bibliography* (Cambridge: Cambridge University Press, 1961).

'War services of members of the Association of American Geographers', *Annals of the Association of American Geographers* 9 (1919), pp. 53–70.

Ward, R. Gerard, 'Oskar Hermann Khristian Spate (1911–2000)', *Australian Geographical Studies* 39/2 (2001), pp. 253–55.

Warkentin, John, 'George Tatham 1907–1987', *The Canadian Geographer* 31/4 (1987), p. 381.

Watson, J. Wreford, 'The sociological aspect of geography', in G. Taylor (ed.), *Geography in the Twentieth Century* (London: Methuen, 1957), pp. 463–99.

Weber, Ronald, *The Midwestern Ascendancy in American Writing* (Bloomington: Indiana University Press, 1992).

Weikart, Richard, 'The origins of Social Darwinism in Germany, 1859–1895', *Journal of the History of Ideas* 54/3 (1993), pp. 469–88.

Wheatley, Michael, *Nationalism and the Irish Party: Provincial Ireland 1910–1916* (Oxford: Oxford University Press, 2005).

Wheeler, James O. and Brunn, Stanley D., 'An urban geographer before his time: C. Warren Thornthwaite's 1930 doctoral dissertation', *Progress in Human Geography* 26/4 (2002), pp. 463–86.

Whitaker, J. Russell, 'Almon Ernest Parkins', *Annals of the Association of American Geographers* 31/1 (1941), pp. 46–50.

Whitbeck, Ray H., 'Geographic influences in the development of New York State', *The Journal of Geography* 9/5 (1911), pp. 119–24.

——, 'Mountains in their influence on man and his activities', *The Journal of Geography* 9/2 (1910), pp. 54–55.

——, 'Response to surroundings – a geographic principle', *The Journal of Geography* 3/9 (1904), pp. 409–12.

——, 'Review of *Influences of Geographic Environment*, by Ellen C. Semple', *Bulletin of the American Geographical Society* 43/12 (1911), pp. 937–39.

——, 'Thirty years of geography in the United States', *The Journal of Geography* 20/4 (1921), 121–28.

Willis, Bailey, *A Yanqui in Patagonia* (Palo Alto, CA: Stanford University Press, 1948).

Williams, Frank E., 'Geographer-envoy from America to Europe', *Annals of the Association of American Geographers* 20/2 (1930), pp. 86–90.

——, 'Ray Hughes Whitbeck: geographer, teacher, and man', *Annals of the Association of American Geographers* 30/3 (1940), pp. 210–18.

Williams, Michael, 'The creation of humanised landscapes', in R. Johnston and M. Williams (eds), *A Century of British Geography* (Oxford: Oxford University Press for the British Academy, 2003), pp. 167–212.

Wilson, Keith M., 'The "Protocols of Zion" and the "Morning Post", 1919–1920', *Patterns of Prejudice* 9/3 (1985), pp. 5–14.

——, *A Study in the History and Politics of* The Morning Post, *1905–1926* (Lewiston: E. Mellen Press, 1990).

Wilson, Shannon H., *Berea College: An Illustrated History* (Lexington: University Press of Kentucky, 2006).

Winsted, Huldah L., 'Geography in American universities', *The Journal of Geography* 10/10 (1912), pp. 309–16.

Wise, Michael J., 'Becoming a geographer around the second world war', *Progress in Human Geography* 25/1 (2001), pp. 112–21.

——, 'A university teacher of geography', *Transactions of the Institute of British Geographers* 66 (1975), pp. 1–16.

Withers, Charles W. J., *Geography, Science and National Identity: Scotland Since 1520* (Cambridge: Cambridge University Press, 2001).

——, 'Mapping the Niger, 1798–1832: trust, testimony and "ocular demonstration" in the late Enlightenment', *Imago Mundi* 56/2 (2004), pp. 170–93.

——, *Placing the Enlightenment: Thinking Geographically About the Age of Reason* (Chicago: University of Chicago Press, 2007).

——, 'Working with old maps: tracing the reception and legacy of Blaeu's 1654 *Atlas Novus*', *Scottish Geographical Journal* 121/3 (2005), pp. 297–310.

Withers, Charles W. J., Finnegan, Diarmid A., and Higgitt, Rebekah, 'Geography's other histories? Geography and science in the British Association for the Advancement of Science, 1831–c.1933', *Transactions of the Institute of British Geographers* 31/4 (2006), pp. 433–51.

Withers, Charles W. J. and Mayhew, Robert J., 'Rethinking "disciplinary" history: geography in British universities, *c.* 1580–1887', *Transactions of the Institute of British Geographers* 27/1 (2002), pp. 11–29.

Wright, John K., 'The American Geographical Society: 1852–1952', *The Scientific Monthly* 74/3 (1952), pp. 121–31.

——, 'British geography and the American Geographical Society, 1851–1951', *The Geographical Journal* 118/2 (1952), pp. 153–67.

——, *Geography in the Making: The American Geographical Society 1851–1951* (New York: American Geographical Society, 1952).

——, *Human Nature in Geography* (Cambridge, MA: Harvard University Press, 1966).

——, 'Miss Semple's "Influences of geographic environment" notes towards a bibliobiography', *Geographical Review* 52/3 (1962), pp. 346–61.

Zelinsky, Wilbur, 'The strange case of the missing female geographer', *The Professional Geographer* 25/2 (1973), pp. 101–05.

——, 'Women in geography: a brief factual account', *The Professional Geographer* 25/2 (1973), pp. 151–65.

Index

Adams, George, 40

Agassiz, Louis, 27

Almagià, Roberto, 84–86, 90

American Geographical Society, 36, 64, 112: and the Bureau of Inquiry for the Peace Terms Commission, 138–39; Cullum Geographical Medal of, 112; Transcontinental Excursion of, 1, 3

American Historical Association, 25, 40–41, 74

American History and its Geographic Conditions (Semple), 33–35, 38, 40, 43, 52, 56, 59, 65, 111, 113, 164

Amundsen, Roald, 109

Anderson, Esther, 143

Anthropogeographie (Ratzel): 14, 21–23, 39, 44, 78, 146

anthropogeography: application of to the United States, 33–35, 59; definition of, 2, 50; illustrated by real-world examples, 29–32, 50, 58, 92–93, 95–97, 110, 115; promotion of as an empirical method, 2, 32, 34, 37, 59, 71, 73, 77–79, 176. *See also* Ratzel, Friedrich

anti-Semitism, 56, 62, 101, 167: and 'The Britons', 101

Appalachian Mountain Club, 112

Aristotle, 17, 150

Association of American Geographers, 27, 36–38, 40–41, 142, 144, 158, 170

Atwood, Wallace, 39, 126–27, 142–43

Baker, John, 162: reading of *Influences*, 161–62

Baker, Oliver, 142

Barrows, Harlan, 37–40, 114–15, 121–22, 132, 135–37, 140, 144–45, 154, 160: promotion of human ecology, 115, 144–45, 160; as teacher, 121

Baugh, Ruth, 143, 151, 153–55, 158, 170: friendship with Semple, 155

Beloit College, 37

Bergson, Henri, 80

Bird, Isabella, 94–95, 102, 105

Boas, Franz, 22–23, 65, 140, 146, 156: reading of *Influences*, 140

Bowen, Emrys, 160

Bowman, Isaiah, 138–39, 141, 149: as director of the American Geographical Society, 138; friendship with Semple, 139

Braithwaite, William, 55

Brigham, Albert, 27, 33–34, 98, 121, 124, 127

British Association for the Advancement of Science, 76

Brown, Ralph, 115

Brunhes, Jean, 77, 117, 157, 168

Buckle, Henry, 13, 21, 68, 77

Burr, George, 40–41

Burrill, Meredith, 143

Cameron, Charlotte, 101, 105

Carlson, Albert: reading of *Influences*, 157

Carney, Frank, 124–28, 134, 172

Central Wesleyan College, 130

Ceylon (Sri Lanka), 68, 96–97

Chamberlain, Houston, 56

Chamberlin, Thomas, 37

Chisholm, George, 76–81, 85, 89, 92, 96, 108, 141–42, 168–70

City College of New York, 138

Clark University, 119, 126, 142–44, 151–57, 159, 165, 170

Close, Charles, 76–80, 89, 177

Colby, Charles, 171